THIRD MILLENNIUM THINKING

THIRD MILLENNIUM THINKING

Creating Sense in a World of Nonsense

Saul Perlmutter
John Campbell
Robert MacCoun

HODDER &
STOUGHTON

First published in Great Britain in 2024 by Hodder & Stoughton Limited
An Hachette UK company

1

A CIP catalogue record for this title is available from the British Library

Hardback ISBN 9781399705493
Trade Paperback ISBN 9781399705509

Typeset in Adobe Garamond Pro by Manipal Technologies Limited

Printed and bound in Great Britain by Clays Ltd, Elcograf S.p.A.

Hodder & Stoughton policy is to use papers that are natural, renewable and recyclable products and made from wood grown in sustainable forests. The logging and manufacturing processes are expected to conform to the environmental regulations of the country of origin.

Hodder & Stoughton Limited
Carmelite House
50 Victoria Embankment
London EC4Y 0DZ

www.hodder.co.uk

To our children, with hope for a world where people can work together in thoughtful deliberation to navigate the third millennium's challenges—and opportunities.

Contents

Part IV
MINDING THE GAPS

Part V
JOINING FORCES

Introduction

In just the past few decades, those of us who live in the internet-connected world have obtained access to a nearly unfathomable amount of information. We can click a link and instantly gain insight into whatever we're curious about, whether it's treatment options for a particular health condition, how to build a solar generator, or the political history of Malta. On the other hand, sometimes there is so much information we don't know how to sort or evaluate it. The social science database ProQuest, for example, boasts of "a growing content collection that now encompasses... 6 billion digital pages and spans six centuries." And that's just old-school, print information! The Internet Archive's Wayback Machine, an archive of websites and other digital artifacts dating back to 1996, hosts almost a trillion pages of digital content, tens of millions of books and audios, and nearly a million software programs.

More and more often, it can be hard to determine what to focus on, let alone how to distinguish what's revelatory and enlightening, in and among all the highly technical, specialized, contradictory, incomplete, out-of-date, biased, or deliberately untrue information we can now access. Was that drug study funded by a pharmaceutical company? Did an AI system invent all those supposedly authentic product reviews? What do those statistics leave out? What does that article even *mean*? It is also increasingly tricky to identify whom to trust for expert guidance in interpreting this information. There are all sorts of people out there who claim expertise—and perhaps *your* favorite experts aren't *my* favorite experts. Experts disagree, or have ulterior motives, or perhaps don't understand the world or "real life" beyond their own narrow perspective. How do we find an expert we can safely trust?

To make a sound decision, take a meaningful action, or solve a problem—whether as individuals, in groups, or as a society—we need first to understand reality. But when reality is not easy to discern, and we're not sure which experts to trust to clarify the matter, we adopt other strategies for navigating the clutter. We "go with our gut"; decide what we "believe" and look for evidence to reaffirm whatever that is; adopt positions based on our affiliations with people we know; even find reassurance in belittling the people who disagree with us. We choose to consult experts who tell us what we like to hear; or bond in shared mistrust of people providing or communicating the information that confuses us, whether they are scientists, scholars, journalists, community leaders, policymakers, or other experts. These coping strategies may help us get by in our personal or professional lives; they may provide a consoling sense of identity or belonging. But they do not actually help us see clearly or make good decisions. And resorting to them can have dangerous social and political consequences.

How can we navigate better—as individuals, and as a society—in this age of informational overwhelm? How do we ward off confusion, avoid mental traps, and sift sense out of nonsense? How do we make decisions and solve problems collaboratively with people who interpret information differently or have different values than we do?

The three of us—a physicist (Saul), a philosopher (John), and a psychologist (Rob)—have been working closely together for nearly a decade on a project to help our students learn to think about big problems and make effective decisions in this "too much information" age. We began our collaboration in 2011, in response to what was already a worrying trend toward no-think, politics-driven decision-making. An issue like raising the national debt ceiling, for example, was being debated that summer as if it were a religious schism, rather than a simple, practical, probably even testable question of what economic approach would work best to improve the country's economic well-being. Most of the arguments both yea and nay betrayed equal disregard for, or ignorance of, the most basic principles of scientific thought. We began to wonder whether it might be possible to first articulate and then teach the principles that would lead to clearer thinking, more rational arguments, and a more fruitful collaborative decision-making process.

The result was a team-taught, multidisciplinary Big Ideas course at UC Berkeley, intended to teach students the whole gamut of ideas, tools, and approaches that natural and social scientists use to understand the world. We also designed the course to show how useful these approaches can be for everybody in day-to-day life, whether working individually or collaboratively, in making reasoned decisions and solving the full range of problems that face us. To our great satisfaction, the course has been both popular and successful, and has since been replicated and adapted by other teachers at a growing number of other universities.[1] Our students appear to rethink their worlds and emerge energized with new ways to approach both personal decision-making and our society's problems. They are better able to investigate their questions, evaluate information and expertise, and work together as members of a group or a society. Inspired by their enthusiasm, we began to think about new ways to share these tools—and this new way of thinking and working together—beyond the classroom, with students and citizens of all ages.

We have become ever more concerned that our society is losing its way, causing suffering—and missing great opportunities—simply because we don't have the tools that could help us make sense of the extraordinary amount of complex, often contradictory information now available to us. Practical problem-solving can come to a standstill when we cannot ascertain the facts of the problems, or, when those problems require communal or political solutions, even agree with others on what those facts are. We humans, who can figure out rocket science and fly to the moon, can't always figure out how to navigate uncertainty and conflicting points of view to make a simple reasonable decision when we need to.

How *did* we fly to the moon? How is it that we have—through centuries of our own efforts as a thinking species—progressively reduced hunger and expanded longevity for an increasingly large fraction of humanity? How did we reach a world where most have access to magical communications capabilities—and seemingly infinite information? Why can't we use whatever it is that got us this far to solve the global problems we face today—problems like the pandemic, climate change, poverty, and so on? Why aren't we able to use the intellectual tools that worked so well in the past?

Part of the problem is that science itself is often a major source of the highly technical, opaque, inconsistent, and contradictory information that has overwhelmed, perplexed, and even angered people. Trust in science has eroded in the recent past.[2] The achievements of science cannot live up to all the utopian expectations those successes have generated. Some scientific achievements have also come with negative social, political, or environmental side effects. For these and other reasons, science has become one of the totems of polarization in political discussions. In short, as science became harder to understand, was connected to undesirable side effects, and subjected to politically partisan critiques, many people lost their trust in scientists and in "science" itself.[3]

But science also has a phenomenal record of providing insight into—if not answers to—the most confounding questions humans have thought to ask. It has helped us to solve puzzles, address problems, and make better lives over millennia. It is a culture of inquiry rooted in the dawn of humankind, with centuries of practice in evaluating conflicting information in a baffling world, and in distinguishing what we know from what we don't. Along the way, scientists have learned from both successes and mistakes, breakthroughs and blunders, to refine the tools with which to address new questions and solve new problems.

Some of these tools are physical objects, like measurement tools and instruments—from the sextant to supercolliders to quantum computers. But others are thinking tools—habits of mind, rubrics, approaches, procedures, standards, ideas, principles, stances. These thinking tools function as intellectual "hacks," enabling scientists to work more efficiently, with greater chances for success, in a world of many languages and cultures, so as to produce more reliable results. They establish parameters for evaluating information and for distinguishing what we know from what we believe; they encourage us to correct for our own blind spots, biases, and limitations, and to persist even when problems seem unsolvable. They also reflect centuries of wisdom about the essential value—even the necessity—of collaboration, particularly with people who see things differently. While science still involves a lot of trial and error, we don't have to start from scratch; and we can avoid making today at least some of the errors we made yesterday.

Scientists have long been guided by these thinking tools, but many of them aren't in common use in other domains. We believe they could and should be; that they have much wider relevance, and could help us in many more realms and situations—wherever people are trying to evaluate information and expertise, make decisions in the face of uncertainty, and solve the problems that affect their lives, whether as individuals, in communities, or at global scale. In fact, we believe that having more and more people become more and more adept at using these tools is essential to human and planetary well-being in the years and centuries ahead. To survive and thrive in the Third Millennium, we need Third Millennium Thinking.

Many of the challenges we face today in our personal, professional, and political lives, from medical issues to business decisions to social and environmental policies, involve reckoning with highly technical scientific information. This book will address what that information means and doesn't mean; and which questions—emotional, moral, philosophical, and spiritual—technical information can and can't answer. But the scientific framework in the book is useful regardless of whether the information we are grappling with, or the problem we seek to solve, is "scientific." Rather, it provides a perspective applicable to our daily lives as human beings interacting with other human beings in a complex, ever-changing world. Does it make sense to go into debt to enroll in that graduate program? Should I sign up for that medical research study on a new protocol for treating pancreatic cancer? What are the most effective interventions for my child's learning disability? Should our town approve the use of an herbicide to deal with an invasive aquatic weed? Should we spend our school's facilities budget installing solar panels? How should our state regulate self-driving cars?

The critical tools of science can help us navigate and make decisions around these complicated issues. You don't have to be a rocket scientist, or a scientist at all, to understand or make use of what science has to offer. What has been missing is a good translation—a clear and concise explanation that expresses the scientific approach in an accessible way, and that illuminates its practical uses in everyday life. That is what we have aimed to provide in this book. To do so, we draw on our three widely disparate areas of expertise:

John draws on philosophy to look for ways the questions and worries that we face today have been addressed in earlier times, and ways they are new or sharpened in our day. He also brings in the many perspectives that he hears from nonscientists, which can reflect how scientific research looks when it is read or viewed in the news. And he has great stories to tell! (Unfortunately, the reader is left to imagine his Scottish accent.) Rob provides a social psychologist's analytic perspective on how people behave. His public policy and law experience complements this expertise with practical examples of societal decision-making in real life. Rob has helped policymakers arrive at decisions on such topics as the repeal of the "Don't Ask Don't Tell" policy on sexual orientation in the military, and legalization of marijuana in California, Washington State, and Vermont, leading to his own fund of stories. Saul has worked with scientists across disciplines on science that can be as intergalactic as the expansion of the universe or as immediate as medical sensors and climate measurements. He aims to humanize this often-alien-seeming world and convey what scientists really think they are doing, so that nonscientists can recognize themselves in his personal stories of science in action. Together we have endeavored to introduce you to the elements of scientific thinking in an entertaining way, using thought experiments that we hope are provocative, and relatable examples from everyday life.

We begin, in part I, by focusing on the culture and tools of science and their practical ability to build trust in a shared understanding of reality that can guide us in our decision-making. Part II offers science's tool kit of probabilistic thinking as a potential superpower we can all use to get the most out of a world full of uncertainties. And part III introduces the "radical can-do stance" that scientific thinking offers us when tackling big, complex, and slow-to-solve problems—our second superpower, if we are allowed two. We offer it along with tricks of the trade that make such can-do solutions possible.

With these tools of scientific thinking in play, we take a sharp narrative turn to the equally challenging task of applying them in the messier process of decision-making, where facts and figures meet values, fears, and goals. Part IV looks at the myriad ways our individual thinking tends to go wrong, and some new and not-so-new techniques to sidestep these mental traps, developed for science but useful for everybody. Finally, in part V, perhaps the biggest question of our day: What ideas have we learned that we can build on to solve problems with others—our partners, our teams, our society, and our world—successfully weaving all the rationality we can muster in with our very human emotions?

The idea that we can develop more such practical, principled ways to bring us together, to join forces, might be the most important key for our collective futures. Today we face possibly catastrophic climate change, the threat of global pandemics, and runaway stratification of wealth. As we solve these possibly existential challenges to our civilization, there may be others: a large asteroid may be on a collision course with Earth; or the ash cloud of the next megavolcano may completely disrupt all air travel and create a global crop failure, followed by a crop blight. But neither the current threats nor the future plausible catastrophic scenarios would feel anywhere near as scary if we were *working together,* even partially, taking full advantage of our best 3MT skills. Together, we can solve big problems!

A word about the term "Third Millennium Thinking," the very title of this book: We intend "3MT" in the most playfully grandiose way, to describe a collection of ideas and approaches that we have seen people beginning to use as we enter the Third Millennium that appear particularly fertile. These continuously improving ideas and approaches come from a variety of sources and traditions, but a current version of scientific thinking is a very big contributor. While many of the ideas will be familiar to at least some readers, we haven't assumed that that's the case, and have aimed to provide a self-contained introduction to each of them in the following chapters. (We cheerfully give permission to skim-read any that you already know.)

In laying these ideas out all in one place, our goal is to make the case that, taken together, they have begun to form a route forward for all of us, in a complex world. We also think that they are just plain useful for day-to-day life, with all the information we need to parse, all the decisions we need to make, and all the planning and collaborating we need to do, as individuals, as parents and family members, and in groups and organizations. But further, we also believe that our own futures depend on teaching these ideas to others—because even we authors are, by ourselves, incapable of consistently avoiding the errors that inspired their development in the first place. Maybe we sidestep the errors just a little more effectively on the days when we are lecturing about them in a classroom, but for our professional research work we depend on a full, strong culture of other researchers who have been scientifically trained to watch for these failings and mental traps; together, we try to keep one another honest. For all the other problems of the world outside of our research work, we need to depend on those of you who, we hope, will learn from this book how to watch out for us and for one another.

Over the past few years, we have all become aware of the shocking degree of polarization in our society, and the surprising interaction between this polarization and our society's often-problematic relationship to science and scientific expertise. If we are to have any hope of finding the practical common plans and common understandings that can move our society ahead together, we need to learn to accept the possibility of errors in our own thinking, and our need for opposing views that help us see where we are going wrong. And we need to understand the source of the disenchantment with and backlash against scientific progress that arose during the end of the Second Millennium and seek to repair it.

No one book and no single approach can heal the rifts. Not all of our polarized disagreements will vanish. But we have to start somewhere. And we believe that one of our more promising starting points is with the culture of science—*if* we begin to borrow its tools, ideas, and processes, and make a Third Millennium shift in our own thinking.

Part I

Getting a Grip on Reality

CHAPTER 1

Decisions, Decisions, Decisions

Say you are on a hiking trip with your friends. Suddenly, you feel intense pressure in your chest and you black out. When you regain consciousness, you're in a hospital. Two young interns, who are the only doctors on duty, are looking at a CT scan, and you can hear them talking. One of two things has gone wrong with your heart. Here's the issue: They can't figure out which. If it's scenario A, then you are going to need invasive heart surgery. They are going to need to cut you open right now to keep you alive even for the next few hours. There's going to be a significant risk of complications, including some that might kill you, but without it you die for sure. However, scenario B seems to them to be equally likely. In that case, all you need at the moment is medication. The medication will keep you going for the next two or three days, giving the doctors plenty of time to conduct further tests and monitoring. But if what is really going on is scenario A, then just giving you the medication will mean that you die.

At this point, the interns realize that you have woken up. They ask you how you would like to proceed. "I have no idea what to decide!" you tell them. "I'm under duress here; just save me." They confer for a moment and then offer you two other ways to make the decision. First, they know you're a big fan of democracy, so you could use a democratic approach: ask everyone in your town—parking attendants, regular citizens, council members—to vote on the decision. Alternatively, they tell you, you can have the most knowledgeable and experienced doctors make the decision.

When we have to make a high-stakes decision about something we aren't expert in, or we simply don't know what the right answer is, the first choice we make has to do with who we are going to consult with, or seek information from, to best inform that decision. In many important situations, such as choosing a political representative, or legalizing marijuana, or selecting land where wind farms will be permitted, the question of what the majority of people think really matters. For these decisions, there is much to be said in favor of a democratic approach. But in this hypothetical medical case, it is hard to imagine many people saying that what they most care about is respecting majority opinion. What really matters is the quality of the decision, and usually, you'll make a better decision if you ask a couple of good doctors than if you take a vote.

We don't all have equal knowledge about everything. Some people know more about history than others do; some know more about automobiles than others; and some know more about medicine. If knowledge is power, then you'd be enfeebling yourself if you cut yourself off from the specialist knowledge that various experts have. One reason for listening to experts is that it empowers you to do what you want to do.

But our need for expertise presents us with three conundrums: First, if we don't have specialized knowledge ourselves, how do we even start to think about what knowledge we need and who is a reliable expert in that domain? Second, assuming we have found reliable expertise, how and when do we appropriately fold in the other key elements of a decision: the values, the emotions, the goals? And third, what makes the decision legitimate and respectful of our personal autonomy—who has the final authority and why? Let's begin to unpack each of these questions.

EXPERTS AND PSEUDOEXPERTISE

Science nowadays is usually complex, and was developed using mathematical models that most people can't understand at all. It takes years of training to understand the math. This can lead some people to unquestioningly follow the advice or mandate of experts ("Well, you're not going to understand the formulas, so just do what they tell you"); on the other hand, some people find it so disempowering to be in a position of ignorance, they seize the opportunity to exercise negative power and refuse to listen to the experts.

This dilemma was particularly pronounced during the Covid-19 pandemic. Scientists gave us a lot of advice—"Don't wear a mask," "Do wear a mask," "Get vaccinated to be safe from Covid," "Get vaccinated to be safer if you do get Covid," and so on. But few of us understood the reasoning behind this advice, or why the advice seemed to keep evolving over time. Few of us could even explain what a virus is, or how exactly these measures might help you not get the virus. "Autonomy" in this situation seemed to come down to this: you could try to isolate yourself from all the conflicting information, or you could try to choose among the experts to find those you trusted most.

Confusion caused by information (of varying quality) overload during the pandemic is one instance of a more general problem: there is a bewildering flood of available information on just about any topic that might be of practical interest. When the topic is one requiring specialized knowledge, how are you to proceed if you want to find the most reliable information? Whom are you to trust? Why should you trust some experts over others?

For practical decisions where you need accurate information, there's a particular consideration: You want to be using a source of information that *works*. Take, for example, agriculture. People have been farming for a long time—about twelve thousand years. And if you're going to farm, there are different ways you might try to find out when the best time is to grow corn, for instance. You might rely on the word of a spiritual leader. Or you might rely on the word of someone who claims to see patterns in the stars that say when corn should be planted. These might work fine in a stable environment; spiritual advisors and astrologers may have tuned their messages to the local conditions over generations. But the scientific approach—experiment and observation—offers a powerful advantage. There may be better seed varieties, better ways to water the crops. Try different things and observe to see which seems to be giving the best results. No matter how strong your faith in your spiritual leader, or in the stars, it is going to be hard to sustain that faith when you see your rival's crops towering feet above yours, and that they are in plenty while you are in famine.

The great thing about science is: *it works*. We hardly need to point out how many ways there are in which science enters our everyday lives, from our use of medicine through the food we eat, the cars we drive, and the internet we rely on. We don't expect much controversy about this. (In fact, one of the problems facing us in knowing what to do with the torrent of information we have is how many wild claims are made about what science can do: from the implantation of microchips into vaccines to the replacement of the sun by electronic lamps. Science has achieved so much, and its power is so taken for granted, that for many people, it's not really a stretch to suppose that it could be doing those unlikely things.)

Science doesn't work by magic; it works by design. Science helps to protect us from being taken in by appealing ideas that don't have any substance to them. We all struggle to overcome our natural biases. But scientists have also developed a range of practices to use in protecting against them. These are "impersonal" techniques, in that when you have to size up evidence, you can just run the process more or less mechanically, without your own psychology coming into play. These techniques are common across the sciences. In different disciplines they're given different names, but they're recognizably the same techniques. You should know about these techniques if you are not already using them. They're science, but they're not rocket science. They're not hard to understand, they don't require math, and they allow you to engage with what scientists say, whether or not you are a scientist yourself.

Take, for example, a scientist in molecular biology whose career has been spent working on specific proteins in the human body. A scientist like this will be able to look at an experimental study in developmental psychology—say, something about how children learn arithmetic—and understand the logic of the experiment's design. That isn't because a career in molecular biology gives you a lot of specialized knowledge about how children learn. It doesn't have to do with mathematical skill; there may not be any complexity involved in the mathematical analysis here. Rather, it's because all experiments, in whatever scientific discipline, are subject to the same kinds of problems, and all scientists have the same set of biases that they need to protect themselves against. It's not hard to learn about these techniques, even if we don't have scientific training. In fact, it's not hard to learn about them even if we don't have much in the way of formal schooling. And these techniques are essential outside the context of academic research, even when you're thinking about very practical topics, like what to feed your child, or whether you should get vaccinated.

With an understanding of these techniques, we're not going to be able to rerun the scientists' experiments for them, or develop the kind of specialized knowledge of a scientific subfield that takes years to achieve. But we're able to assess whether we're looking at something that is honest work, that might take us closer to the truth, or merely some brightly colored tale that plays to our prejudices. We're able to distinguish between expertise and pseudoexperts. And equipping ourselves to do this—articulating the techniques and tools of scientific thinking—is one of the main goals of this book.

MAKING THE CALL ON VALUES

In most circumstances, however, facts are not the only factors we need to consider. Indeed, for some decisions—Who is funnier: Charlie Chaplin or the Marx Brothers? Which is better: the red sauce or the green sauce?—you may not feel you need any outside facts at all. But most of the time, while you need facts, they are not all you care about. Values, ethics, fears, and goals are often significant factors in decision-making.

Even in medical cases, two people faced with the same knowledge might make different treatment choices. Suppose we go back to the scenario where you wake up in the hospital bed. You have to make an assessment about where the evidence points. But you also have to weigh up what matters to you in this situation. Different people might weigh risk differently. You might think: Well, there are even odds that I'll be fine if I just take the medication, but I can't take the risk that I might die. So I'll go for the surgery. There might be complications, but it's tried and tested and unlikely to kill me. Or if you're a bit bolder, you might say to yourself: I don't want to go through the agony of surgery and surgical recovery if the chances of survival aren't superior, so I'll take my chances with the risk of death and choose to take the medication. What value to put on the risk is really up to you. Doctors can tell you how big the risk is, but not how much it should matter to you.

This kind of problem is especially poignant for the parents of a child who's been diagnosed with, for instance, cancer. You might find entirely unacceptable the idea of your child being given a radical treatment that has a significant risk of a disastrous outcome but that will work if the odds go your way. Or you might insist on the radical treatment, unable to bear the notion of the disease progressing in your child. What values you put on the risks are up to you; experts can't advise you. It's impossible for anyone but you to decide whether the risk of your child dying from cancer outweighs the risk of your child living but suffering some permanent ill effects of the treatment. There's no complicated mathematical formula or scientific experiment that will tell you how to weigh those factors.

There's a bundle of considerations here that we might construe as "values" and contrast with the fact-finding that scientists are trying to do. You might draw your values from your own family background, or a religious community you belong to, or just the people around you and the books you read. People differ in how they define a particular value, and most people embrace a set of values that may not be entirely consistent. All these factors might influence how much weight you put on the risk of various possible bad outcomes when you're making a decision, and how important you think the various benefits that might come to you are.

There aren't "experts" on how to evaluate these matters in the way there are for the possible risks or benefits of a vaccination, for example. But there are, of course, people who have thought a lot about various moral problems, particularly problems that come up often in real life, and who are familiar with all the various considerations that can generally come up on one side or another. There are good reasons why hospitals and universities frequently employ such people to help with practical decision-making. Many of us have people we particularly want to talk to when something morally difficult comes up—a parent or partner or pastor or old friend. But there aren't panels of experts universally recognized as authorities on making value decisions, the way there are panels of experts on things like the health implications of tobacco use.

Things get more complicated when a group or community is engaged in a decision process. Not only might it be hard to agree about facts (although if you are equipped with the tools we've just introduced and will discuss further in later chapters, you may be able to identify reliable sources of information), but there might be people with different, and even conflicting, values to consider. We will focus on this challenge later in the book.

EXPERTISE AND AUTHORITY

So we need to have sources of trusted, factual information, and we need to consider that information in the context of our values as we think through alternatives and try to understand potential outcomes of our actions. But in the end, who has the authority to make decisions?

In most societies today, there is a presumption that individuals have the right to make the decisions that affect them. But have you ever considered why *you,* in particular, should have any rights over the decisions that most affect you? This question goes to the center of a lot of current disputes.

Suppose that what we all want, you included, is for things to go well for you. Most of us can remember times when we got things wrong, and did things we regret, or made decisions that led to things going badly for us. We're not always particularly expert on what's going to make things go well for us—not only when it comes to complicated medical decisions, but in many other realms as well. If what you want is simply for things to go well for you, maybe you should just turn over all your decision-making to experts.

To most of us, this will seem like a nightmare idea. The idea of a society run by experts who decide what and when we should eat, what kinds of medications we should take, and what kinds of medical procedures we should have; who decide what jobs we should go for and what social groups we should belong to, what kinds of exercise and romantic partners we should have, seems like a kind of hell, even if they got it "right." We want to keep the right to toss aside the advice of "experts." But isn't this just irrational? You might argue—and you might be right—that the "experts" sometimes get it wrong, and anyway may not understand your best interests as you see them. That is certainly a possibility. But it's also possible that you are subject to self-destructive impulses that are just as bad; that you'd be much better off if the experts were the only people who had the right to make decisions affecting you. So why does it stick in the throat to give up power in this way?

The natural answer is, you've been brought up—at least in a democratic society—to see yourself as a free person, with your own rights and responsibilities. You expect your freedom to be acknowledged and respected. Consider again the case where you're coming around after a heart attack and the question is what to do. At the end of the day, it's up to you to size up the situation and make the call. You can't just be rolled over by a bunch of experts who tell you they know what's best for you. In the end, *it's your call.*

But in many important decisions, it's not your own personal welfare that's at stake—it is the welfare of another person who does not have the capacity to make the decision. Suppose Grandma is dying, and unlikely to recover full functioning; she's unconscious and on life support at this moment. She's given you the authority to pull the plug. When, if ever, should you do it? Well, an AI enthusiast might say, give the decision over to a machine that will take into account the best available medical and statistical evidence about likely outcomes. That might involve much more complicated calculations than you could ever do yourself, based on all sorts of details concerning your grandmother and current medical knowledge. The trouble is that in this kind of situation, you have to take ownership of the decision—your grandmother asked you to. You have to make the call. You can't hand it off to the machine. It's not good enough to say, "Well, the machine said to let Grandma go so I pulled the plug." Maybe the machine could supply you with relevant arguments and considerations to take into account, but you have to understand them and weigh them and make the final decision. Being a free, autonomous person means that you have to make the values call regarding this person who has placed themselves in your care by yourself, even if you listen to a lot of other people, or to machines, before you do so, and even if you follow their advice.

Making a choice for yourself seems different from making a choice about an aged relative, or a child with cancer, or a baby who can't speak for themselves, or choices about animals, or trees, or inanimate objects. Consider animal husbandry. If you're running a farm and want your livestock to do well, there are centuries of lore and science about how they should be managed, and we expect farmers and ranchers to make use of that information in running their operations. We're not bothered by the fact that farmers and ranchers make these decisions without consulting the cattle; rightly or wrongly, we think of ourselves as *free* in a way that animals are not. There is a huge difference between a scientist giving us advice on human affairs and a farm manager deciding whether, for example, the livestock should be vaccinated. We recognize other people as autonomous individuals, and we expect them to extend the same respect to us. (Part of what is so agonizing about making decisions for an aged relative or a young child is that we wish we *could* consult the patient, even though we can't, because we believe it is their right, not ours; we tend not to be troubled in this way about animals.)

In some cases, we need to engage in decision-making not just for an individual who cannot decide for themselves, but as part of a group or society that shares, collectively, the position of being the community affected by the outcome of the decision. Although in the heart attack example you are likely to have rejected the democratic decision-making process, voting is one way of including members of a community in a decision that will impact them. In engaging in collective decision-making, we may have to consider multiple and conflicting interests and values, as well as different opinions about what facts are reliable and which experts are trustworthy. Later in this book we will discuss some methods of getting groups not only to evaluate information collaboratively but also to engage in thoughtful consideration and weighing of one another's values, enhancing what we usually consider to be a democratic process, voting.

There are also some special cases where the authority to make decisions may not lie with the people who are most affected. This includes not only the situations we have discussed in which people are incapacitated, but also situations in which the effects of what someone may consider to be a personal decision are likely to reverberate beyond them as individuals. It is for this reason that motorcycle helmets are required, for example, or that public health agencies are given the authority to decide when to close schools during a pandemic. But in most cases, we assume that individuals or communities are rights holders over decisions that will impact them.

FAILURE MODES

Good decision-making, then, depends on all three ingredients that we have discussed: accurate information from reliable experts, a careful consideration of values, and a structure that places the authority to make the decision in the hands of those who will be affected. If any one of these ingredients is dramatically out of balance with the others, we see clear failure modes—we know that something has gone pretty far wrong.

For example, what happens if we overweigh the role of expertise in our decision-making? Some political philosophers have recently been talking about an extreme version of this: the idea of an "epistocracy"—a society where you need some particular educational qualifications, or knowledge, in order to have any kind of vote at all. Perhaps only people who've graduated from high school should be allowed to vote; or only people who've graduated from college. Or maybe everyone gets to vote, but the better educated you are, the more votes you get.[4]

Clearly, whatever the advantages, there are certainly troubling features about an epistocracy—as we will discuss later in the book. How *can* we work with scientists without giving over too much authority to them? Scientists are not in the position of farm managers, with respect to the rest of us, so we don't want them to treat us like sheep. We don't want scientists to exert any control over us that we haven't given them. We expect to take ownership of the evaluative aspects of our decisions. And if scientists want to influence our choices, they need to *persuade* us. Scientists can explain to us what facts they think they've found, and show the practices that they applied in their work to ensure that they obtained an unbiased result, so that we can make up our minds about whether that result is convincing.

This means that everybody, scientists and nonscientists, must have some understanding of the techniques that scientists use to come to their conclusions—unsurprisingly, the same understanding that we discussed as important in our ability to choose good experts. And, as we have said, these are not hidden mysteries; we can all learn them, and this is one of the main goals of this book.

What happens if we overweigh the role of autonomy in our decision-making? This failure mode has happened when the delicate balance between the elements of a decision has been taken to be an either-or choice: "You can either give up your freedom to the technocrats, or you can keep your freedom, reject their so-called expertise, and 'do your own research.'" This might mean, for example, spending a couple of hundred hours looking at YouTube videos and working with what "rings true" to you. Of course, the trouble with this is that what rings true to us is just as likely to be a deadly mistake as it is to be right. We have biases that incline us to believe what's being said by particularly charismatic people, for example, or to believe stories that echo prejudices we already have, or that demonize people we don't like. (We'll discuss biases in more detail later in the book.) Our blindness to our own biases means that we are vulnerable to errors, even errors that might kill us, when we're using our own common sense to determine what "rings true." We're in the position of an army under attack whose radar has been disabled. We simply don't know what we have to guard against.

A different sort of failure mode can occur if we misunderstand how to ensure that our collective and individual values are properly weighed against the expertise in a decision-making process, and we insist that the relevant scientists stay out of the values discussion. Of course, we really do want the scientists working on a particular problem to also reflect on how their research findings might be used, and—as in the case of the atom bomb, for example—whether they ought to be used at all. In fact, we hope that a good scientific education encourages such ethical thinking. We don't want scientists to be working on human genome editing or reading thoughts in human brains without considering the consequences, both good and bad. So a more nuanced statement of our goal is that we want to be able to separate out the scientists' factual findings from their best wisdom concerning the values at stake, since we are looking for them to be *experts* on the facts and then *participants* in the values discussion. An expert whom you can trust should be able to help distinguish these two roles when advising us.

These three examples of failure modes are clearly not exhaustive—there are all sorts of ways that this delicate balancing act of expertise, values, and autonomy can go wrong. Part of our job as decision-makers, individually and collectively, is to keep a watchful eye on this balance and on the processes that lead to a decision. Interestingly, here, too, there is a role for expertise. In particular, the expertise needed to understand the (sometimes democratic) process used to make societal decisions, and the expertise needed to explore the consequences for society of a proposed policy, are often themselves another form of scientific thinking: Ideas and results from social science, as we'll see throughout the book, can be incredibly helpful in determining how to go about joint decision-making. We can improve the ways in which we, as a society, make decisions, so that everyone's reasoning and preferences are given due weight.

Such expertise can also be particularly important in helping to recognize values and goals that we have previously only vaguely considered, but that appear helpful—and necessary—to include in the decision-making once they are articulated. For example, it has been useful to bring out the question of the different time scales on which a societal policy will have its effect, and how much weight we give to the immediate interests of the current population relative to those of the population thirty years (or even thirty generations) from now.

In the end, all of these decisions, from individual to societal, are bets that we place. We rarely are guaranteed that we have made the right choice. This aspect of decision-making can also benefit from scientific thinking approaches that we will be discussing in later chapters, in particular the techniques of "probabilistic thinking."

Everything we've discussed in this chapter also depends on the idea that there's a single reality, the same for all, and that science can show us the way in exploring how things are. But why should we think that science tells us anything about a world "out there" that's the same for all of us? Why should we think that the surprising world science tells us about—with its tiny particles and forces, distant galaxies, electromagnetic radiation, hidden motives, and sudden changes of blood flow in the brain—is really there, and there for us all? If we don't have a shared world, joint decision-making becomes impossible. This is the big topic of the next chapter.

CHAPTER 2

Instruments and Reality

Everyone knows there are partisan disputes over scientific questions. For example, in the US, people on the right tend to think there's not much risk to humans from climate change; people on the left tend to think there's a lot of danger. People on the right tend to think that relaxing the regulations on private gun ownership doesn't lead to increases in crime; people on the left tend to think it does.

It's natural to think that this must be because people on one side or the other don't understand science well enough. People on your side understand science, those on the other side don't. If that were all there was to it, then maybe the way to bring us together on questions like these would be to increase the level of scientific comprehension all around. But, in fact, social scientists have found that there are scientifically literate people on both ends of the political spectrum, and that informing people of "the facts" is rarely sufficient to defuse political disagreements.

As we will examine more fully later in the book, we often pick sides on these hot topics based not on the evidence we've seen but on the identities we hold. Indeed, scientific sophistication can be misused as a way to *weaponize the evidence*—to find ways of using the evidence to promote your preferred beliefs. If you're on the left, and all your friends think intelligent design is a crazy idea, then you could pay a significant social price if you say you think there might be something to it. Knowing something about the subject allows you to marshal the evidence so you can confirm the belief that your social group already holds. If you're on the left, and all your friends think that weakly regulated gun ownership leads to high crime and firearm death rates, then you will have a cost to pay if you want to explore the idea that gun ownership deters crime or that gun regulation is ineffectual. And if you're on the right, vice versa. A bit of knowledge, then, simply provides a way to defend the views endorsed by your group.

Well, you might say, how much of a problem is this? At an individual level, you will be socially more successful, you will get on better, you will have an easier life, if you just reflect the opinions of your chosen group. If each of you is grooving along in the church of your choice, then who cares which one is "right"? And anyway, is there really some notion of "getting it right" that goes beyond falling into step with one's immediate community, respecting the power structures that keep your life in place? Is it possible that there's no such thing as a single "truth" that, whatever our partisan loyalties, we all need to agree on?

RIGHT, WRONG, AND SCIENTIFIC ASPIRATION

In fact, people very seldom accept this picture of truth as something that can be different for different people. Whatever side you're on, you're going to feel that people on the other side are getting it wrong; that they're making serious and perhaps dangerous mistakes. People don't regard the differences of opinion on these kinds of cases, such as the implications of gun ownership, in the same way we'd regard differences about what songs we liked, or what kind of pizza to order.

Although some academics in humanities departments refer to science itself as just one set of power structures among many, most scientists are on board with the rest of us in thinking that there is a right and wrong to be had about basic facts in the world. In fact, they take it that the whole point of science is to find out what's objectively right or wrong. Scientists do try to establish that there's a world "out there" that they're talking about, a world where right and wrong—fact and fiction—are independent of our power structures, and independent of our wishes for what reality consists of.

Look, for example, at the procedures scientists actually use in their work. We'll review many of them in coming chapters (and a famous iconic example later in this chapter), and, as we'll see, these procedures don't fit well with the idea that scientists are trying to achieve agreement by bullying. In fact, the way they work has generally been the opposite of a power group insisting on its dogmas. Rather, the authority of science comes from a relentless self-questioning. Scientists will, in fact, often take it as a requirement that for any view they're considering, there's a clear way that view could be shown to be false. And if there's no way a view could be shown to be wrong, scientists will be all the more dubious that the view is right. Many of the great breakthroughs that scientists celebrate happened because someone showed that an idea accepted by the leaders of their subject simply couldn't be right.

The ideas that survive this questioning process have an authority, but not because someone powerful has insisted on them. It's quite the opposite of a cult that threatens with fire or oblivion those who question its beliefs. In science, the questioning is welcomed—which illustrates one way that science is indeed essentially a social phenomenon, but a truth-hunting, collaborative social phenomenon, not a coercive one. The process of challenging and questioning here, like the process of bringing up a child, requires a whole community.

Now, at this point we should really stop and be clear about something that will come up again and again throughout this book. The concepts, principles, and operating methods that we are describing as essential to Third Millennium Thinking are the hard-won current best practices of science. They are what scientists aspire to do all the time, or improve on, and often actually succeed in implementing. But science is a human endeavor. We can all find examples where individuals and organizations fail at these ever-improving best practices, and even examples of where an entire subfield of science fails. Sometimes these failures are a consequence of misunderstanding the practices, and sometimes they are due to bad motives or ends-justifying-means thinking. However, when scientists discuss these examples, they are not proud of them; they quickly recognize them as failures. What we are promoting in this book are the ever-improving *aspirational* aspects of scientific thinking that lead to our best social capabilities. Scientists don't always live up to these aspirations, but we've learned that science makes more progress when they do.

Moreover, even when all is going well and best practices are followed, science is still a rough-and-ready business in its attempts to discern the reality "out there." There is a truth out there, but the theories and models we humans have developed over the centuries to reveal that truth are usually, at best, approximations of it. We accept that the models we have are usually incomplete; they are usually at best rough guides to what to expect. Sometimes, we find that we have a number of different models or theories about what's going on in a particular area, and we use whichever one works most effectively for the purposes at hand.

Over time we develop our theories and models so that they approximate the truth better. We can and often have reached enough accuracy in our scientific picture of what's going on for us to achieve startling successes based on that picture, a point that hardly needs belaboring. People nowadays tend to be as surprised by the limitations of current science and technology as they are by its manifold successes, which are often taken for granted.

STEPS TOWARD A SHARED REALITY

But that raises the question, how are we to reach agreement about what's "out there" in reality, if both sides are capable of weaponizing the evidence to fit with the ideas they already have? How do we achieve a shared understanding of what a reality looks like?

Maybe our strongest sense of a reality out there comes from our sense of touch: If you pound your hand on the table, or tap it with your finger, or crash into it when you cross the room in the dark, you are pretty convinced that the table is there and real. So let's begin by considering that our most promising path to a shared reality is the one that involves our touching, feeling, holding, prodding, or pushing things, and seeing the response. We can do this without too much risk of conflict: we don't have partisan disputes about whether the table is there.

But physically touching something is not the only way to gain a sense of its reality. Most of us are also willing to accept the reality of something we can prod with a stick. Similarly, while a standard of proof is that you "can see it with your own eyes," we do allow for the use of corrective lenses in order to see it more clearly. We might even be willing to look at a small bug using a magnifying glass and still have that same strong conviction that we are in touch with the reality that we have when we look at something directly, without a magnifier.

And over the years, we're getting better and better at using other, more and more sophisticated intermediaries, well beyond eyeglasses or a magnifying glass, that still leave us with an equally strong feeling that we are encountering reality. These intermediaries include instruments that you can now find in your own pocket if you're carrying around a smartphone. In fact, we can now "see," in what feels like a much more direct, interactive way, things that ten years ago nobody could have seen unless they went to a fancy laboratory—and that a hundred years ago nobody could have seen anywhere at all. We're getting to play a lot more with "experiencing reality" than we used to be able to.

For a pretty example, let's turn to our sense of hearing. You can get an app for your phone that turns the phone into a sound analyzer—a "spectrograph"—that allows you to visualize properties of sounds you make when you sing, whistle, play an instrument, or make a noise. The graph below gives an example of what such an app shows. When you whistle a note, you see a line on the screen; when you whistle a higher pitch, you see the line on the screen move higher. It's quite surprising that when you sing what you think of as a single note, you see on your screen what looks like a chord made up of a lot of different lines. When you sing a higher note, all the lines move up together. When you do this, it feels pretty viscerally convincing that when you sing, you're not singing a single note. You're actually singing a chord made up of all these "overtone" notes, as we call these higher pitches that are mixed in with the pitch you're trying to sing.

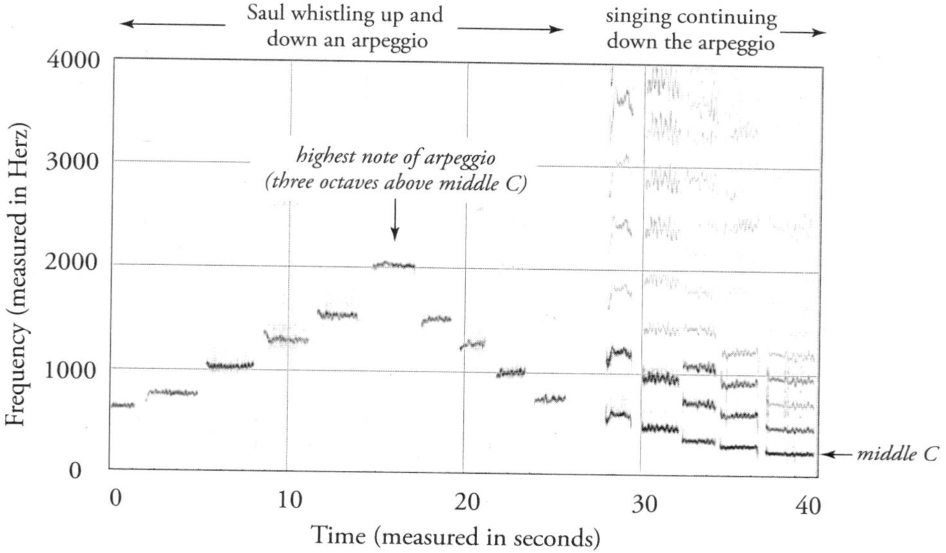

You may also discover that when you sing different vowel sounds, you get different numbers of those overtone lines. If you sing "ah," you get lots and lots of overtones; if you sing "oh" you get fewer; if you sing "ee" you get even fewer. After spending a little time playing with this app, you start to feel that this is the reality of sound. Even if you don't know the theory of it, you start to think of the world of sound in a different way, because you've been playing with it directly. (Our hearing system treats all of these simultaneous, related notes—all of these overtones—as one pitch, but with different "timbres" depending on what mixture of simultaneous notes it detects. This is one of the ways that we tell the difference between a violin, a flute, and a tenor voice, all producing the same pitch. Even without a spectrograph, violins, flutes, and tenors also look different!)

Not every measurement instrument inspires us with this same feeling that we have learned something about reality. What pushing your chair, hitting the table, and singing in the spectrograph have in common is that they produce a feeling of interactive exploration—a term used by the philosopher of science Ian Hacking. The idea behind interactive exploration is that we are more likely to feel confident in the reality of something we experience if the thing we are experiencing changes in response to some input of ours. For example, the billiard ball rolls when you prod it with the cue, so you start to believe that that round image you see over there is an actual hard object with weight in the physical world. The spectrograph display is one step more removed from direct experience, but it, too, responds to your playing with it, by showing a line that goes up and down as you whistle higher or lower, and several lines when two people whistle at the same time. In fact, it even makes you start to believe the reality of things you didn't know ahead of time—for instance, that your voice's timbre reflects several different pitches being produced at the same time.

Once you have started thinking about how your sense of reality is sharpened by interactive exploration, it becomes interesting (and, as we will see, telling) to look at a range of examples of senses being enhanced by instrumentation in which different degrees of interaction are possible. Let's begin with a couple more of these science-y examples and then wander over to examples from day-to-day experience.

After playing with that sound spectrograph app, there is a natural follow-on example that is particularly low-tech—no smartphone involved! Here's how you can do it at home. You need a window that has direct sunlight and a sheet of cardboard with a pinhole in it that you can use to put across the window so only a single beam of sunlight comes through.

You then put a prism in the path of the sunbeam, and see that the sunlight gets spread out into a rainbow of colored light. When you shine the light from an LED flashlight through the prism, you see that only a few lines of color emerge, rather than the full rainbow that sunshine produced. If you can find an old fluorescent or incandescent bulb and shine it through the prism, you find that different color lines emerge from the prism, but, again, not the whole rainbow. We call all these kinds of light white light, but they seem to have different component colors. What can one make of this, and does the prism exercise feel as interactive as the sound spectrograph exercise?

Well, in some ways, this prism examination of light feels quite similar to the spectrograph examination of sound. Just as the spectrograph illustrated that the sound we hear appeared to be made up of many different pitches all heard at the same time, so a prism enabled us to see that white light is made up of a lot of different colors that combine to produce what we see in everyday life. And just as we saw on the spectrograph that a whistle produces a very clean sound, with just one pitch, we can also see that LED light passed through a prism doesn't break into as many colors as sunlight does. As you play with different light sources, you might start to convince yourself that what your eyes are only able to see as white light is actually a "chord" of many different colors combined, the same way the sound of your voice singing one pitch turned out to be made up of many different notes combined. And you might begin to accept that in this way, light is a little bit different from what our eyes alone can tell us.

Comparing these two examples of interactive exploration, sound and light, you might feel like it's great to work with a simple glass prism, as opposed to a phone app. The prism is an object that you can touch. Your mind doesn't start imagining that there may be things you don't understand about the app, or behind-the-scenes machinations fooling you. You're pretty convinced that the light that comes out of the prism is simply a modification of the light that went into the prism. And given the simplicity of a glass prism, you are also sure that the output of the prism is not something that a clever computer programmer is showing you, as opposed to an actual representation of reality.

But there is something that may not be quite as satisfying about this interactive exploration of light with a prism compared to the exploration of sound with a phone app. We can't play with the variety of white light sources to the same extent that we can play with the sources of sound. If we could just shine laser beams from our eyes and change their colors at will—the way we can sing different pitches at will—maybe our brains would be just a little more deeply convinced of the reality of the color composition of white light? In the end, this is the difference between a more interactive exploration of reality and a somewhat less interactive one.

Nevertheless, it could have been worse! Compare these examples to what happens when we want to know how stale the air is in the room where we are sitting—and breathing. In recent years, we have learned that this is an important question. As you breathe in oxygen and breathe out carbon dioxide, the air becomes more "stale," and it becomes hard for your brain to get the oxygen it needs to think clearly. Generally, this doesn't matter, since there is lots of air in a reasonably sized room, and some fresh air with fresh oxygen is leaking in all the time. But if you are in a room without much airflow together with many other people—say in a university lecture room during an hour-long class—then the carbon dioxide fraction can build up while the oxygen fraction drops. Researchers have done studies of how people perform on cognitive tests at different levels of carbon dioxide concentration.[5] People do well when the concentration is 800 and below. They do less well when it is about 1,000. They do pretty poorly on the cognitive tests when the concentration is 1,200, which is close to what levels would be in many poorly ventilated lecture rooms toward the end of an hour.

Nowadays, it is possible to buy a small sensor that shows the carbon dioxide concentration in the air (along with the temperature and the humidity—it's a bargain). Our carbon dioxide sensor does not give us quite the amount of interactivity that some instruments give us. You might be able to get the carbon dioxide number on the display to go higher for a minute or so by breathing into its intake vent, but that's about it. You can't really play with this instrument interactively—and it's hard to know what it is responding to, since we can't otherwise tell how much carbon dioxide is around, the way we can recognize different musical notes and musical instruments when we play with the sound spectrograph. If I tell you that there is more carbon dioxide in the room than when you arrived, and that your brain function is worsening as a result, it's really hard to tell for yourself. (And of course, the less well your brain functions, the harder it is to tell.)

The upshot is that it's hard to get a sense of "how real" the carbon dioxide concentration number is. We're taking it on faith. For all we know, the readings put out by the sensor are actually caused by aliens sending us transmissions from space, and the aliens presumably give us higher ratings when they see somebody breathe into the intake vent. All right, we probably wouldn't fall for something so bizarre, but it should be clear that the carbon dioxide sensor doesn't have the same immediate connection to reality provided by some of the tools we have just been discussing.

One last example is probably worth considering, since it showcases an extremely simple tool, not high-tech at all, that helps us extend our senses: pencil and paper. During a cholera epidemic in London in 1854, nobody knew what was causing it, so John Snow started plotting cholera deaths on a map. He put a little dot every place where someone died of cholera, and he started to see that they were all centered around a certain area, and that the farther away you got from that hot spot, the fewer deaths there were. He eventually realized that in the center of the hot spot were a well and a pump that people were using for water, and that that particular well had gotten contaminated. He took the handle off the pump so nobody could get water from it, and cholera deaths decreased.

You can think of this as an example of the kind of interactive testing of one's understanding of reality that we're talking about. It is also an example of a low-tech scientific technique that we use to shore up one of our weaknesses. We are not very good at remembering where a lot of events happen over time, so people in London didn't see the pattern of the locations of the deaths. But with pencil and paper, we can record events and thereby get over the limits of human memory and our brain's limited ability to picture lots of spatial data. These tools allow us to do something that we otherwise wouldn't be able to do in terms of sensing what's happening in reality.

Since John Snow's time, we've been inventing more and more ways of extending our senses and overcoming our limits, and so more of the world's reality is becoming available to us. And the instruments that allow us to play interactively with whatever phenomenon is being detected help us build a stronger sense that we're not just making up scientific concepts and names out of thin air; they feel tangible, real, whatever names we give them. We don't find ourselves in disputes about the reality of these phenomena.

REALITY BEYOND THE SENSES: TESTING THE TOOLS

All animals, including humans, have limitations in how they find out about the world. The kinds of instruments we've been describing correct for weaknesses in our perceptions of the world. The most basic weakness we have is that our perceptions don't tell us everything about what's going on. So we need corrective devices. Some of us need spectacles. To see very distant things, like distant galaxies or planets, we use telescopes; to see very small things, like cells, we use microscopes. It's hard for many of us to hear the difference between a single tone and a chord: the sound analyzers we talked about let us break down complex sounds into their constituents, in a way most of us couldn't do unaided. We usually see daylight as undifferentiated white light: it takes the prisms to let us analyze the complexity of daylight, to see that it is made up of rays of different colors.

But the acceptance of the instruments we use in analyzing our surroundings is hard-won. Consider electricity: To find out about electrical currents, we use various measuring instruments: voltmeters, ammeters, and so on. These instruments tend to be familiar, so nowadays we take it for granted that the instrument does what it says on the tin. "It says 'voltmeter,' so I guess it's measuring volts," we say. We tend to forget that the very existence of these instruments was an achievement. How do people know what exactly these devices are measuring, when the instruments themselves seem to be the only ways we have of finding out what's going on?

Let's look at a historical example: Galileo's first use of a telescope, in 1609. When Galileo first pointed a telescope at the night sky, he made many basic astronomical observations he couldn't have made without the telescope. For example, he found that the moons of Jupiter were going around Jupiter itself. Up until then, the truth as revealed in the Bible had been that all heavenly bodies revolved around the earth. So Galileo's observations were certainly the locus of a bitter partisan dispute.

His critics were quick to argue that all Galileo had shown was that if you put lenses together in a tube in the right way, you would get really peculiar spots before your eyes. The telescope, they argued, didn't have any validity as a way of finding out about things "out there."[6] So how would you go about arguing that the telescope is a way of finding out about reality?

As we've noted, the approach of science to this kind of question is quite different from a power-based approach, such as that of cults or, as in this case, some religions. This particular historical example shows the distinction quite vividly. Cults and authority-based religions generally insist on the certainty of the truths they've found, and, historically, have often tried to enforce agreement with them, using one or another type of coercion. For example, on June 22, 1633, the Catholic Church showed Galileo the instruments of torture used by the Inquisition, to encourage him to agree that all heavenly bodies revolved around the earth.

Scientists, in contrast, typically don't use coercion. They embrace the skepticism. They ask whether they might not, after all, have gotten it all wrong. Critically, they use impersonal techniques, applying well-worked-out rules, to check the theories they use. We'll look at many of these techniques in detail in later chapters. But the kinds of questions they're asking to discern whether a particular observation of reality is accurate or not are familiar. How many observations were made? Did different observers all get the same results? Do we know how the instruments we're using work? Is there reason to think those instruments are, or aren't, identifying the phenomenon we're concerned with? And so on.

So, the fact that, in this historical case, Galileo could repeatedly get the observations he did was forceful evidence that they were reliable. There was, of course, a serious problem in communicating these findings to others, because telescopes were originally being built from pairs of spectacle lenses, and they were just not of high-enough quality to sustain accurate stable vision at long distances. (Vision anyway was seen as more prone to illusions than touch.) So Galileo developed techniques that allowed him to make much more powerful and accurate images. But then to convince people, enough of these instruments had to be made available to other researchers. After a while, this happened—and then anyone could begin to test what the telescope seems to reveal by, for example, looking at objects close enough to make changes to, like the cows one field over. But this was only the entryway to Galileo's overarching project, which was to show that the heavens and the earth, though they seem so different, contain pretty much the same kinds of stuff, governed by the same mechanical principles. It had for centuries been thought that the glory of the heavens is quite different to the glory of the earth. Galileo's work in shaping up the idea that a principle of inertia, for example, could apply to both the celestial bodies and the earth was fundamental to Newton's finding a single set of equations governing matter of any sort, heavenly or earthly.

BUILDING WITH WHAT WE KNOW: THE RAFT AND THE PYRAMID

There isn't, then, one single "magic bullet" that validates Galileo's conclusions, but rather, a web of considerations that reinforce one another, each of which can be independently tested ("Have we got it right about refraction?" "Do different observers all get the same results?" and so on). We can explain this in terms of two classic pictures of the structure of science: the raft and the pyramid. Here's how it goes.

When adventurer and ethnographer Thor Heyerdahl took his balsa raft, the *Kon-Tiki,* on its trip from Peru to Polynesia in 1947, his crew predicted that the balsa logs from which the raft was built could become waterlogged on the journey. So they took with them spare balsa logs. That way, if any one of the logs from which the raft was built became waterlogged, and so unusable for flotation, they could strip it out and replace it with one of the fresh logs stored on board. But what they couldn't do, of course, was to strip out and replace all the logs simultaneously. The moment they stripped out a number of the logs, the whole raft would collapse, and they would drown.[7]

This image of the raft works quite well as a metaphor for the crisscrossing pattern of justification that we use to demonstrate that an instrument, like the telescope, works and is giving us the information we are counting on it to give us. Suppose you tried to suspend belief in everything: You don't accept anything at all of current knowledge, and then try to reconstruct all that we do know from scratch. That means throwing out everything from knowing how to tell if someone's illness can be cured by antibiotics, to knowing whether spots mean measles, to knowing the patterns of movement in the night sky, and then justifying all that we believe from scratch, including, for example, which vaccines will work on which diseases. That would be like throwing away all our logs to rebuild the raft from the beginning: We wouldn't be left with enough to work with. We would drown. But what we can do is test each proposition individually, while keeping steady most of the background, and toss out and replace ideas that don't pass muster. Given most of our current background of medical knowledge, for example, we can go back and review whether a particular vaccine is really protecting against a particular illness. And similarly, for each medical proposition we believe, we can, holding the rest of the background constant, review and assess whether it's right.

An alternative image to the raft is the pyramid. Here, the idea is that we view science as organized in layers, with higher layers depending on the lower layers. This is quite different from the raft metaphor, where we don't think of some groups of logs as being more fundamental than others. In this picture, there's a base level of scientific beliefs that support all the rest, and we couldn't question them without bringing the whole thing down.

Neither of those images is entirely satisfactory. But the image of the raft better fits the cautious, skeptical, provisional nature of the approach to science that most people currently take. There are no propositions that can't be questioned and maybe thrown out and replaced. But when we're doing that, we should keep in mind the "local" character of the questioning we're doing: We can ask and answer questions about the justification of any one proposition only by holding some other things fixed, just as we can check and replace logs only one at a time. But every proposition can, in turn, be challenged, just as any of the logs can, one at a time, be checked and replaced.

The raft metaphor also captures another key part of the story. Each element of our scientific understanding, each log in the raft, only gets its strength by relying on all of the other scientific-element logs that it is connected to. We trust one bit of science because there are many other bits that, together, support it. In this sense, we are triangulating with a number of different pieces of evidence to trust any other piece of evidence; that is how the scientific raft functions. As we will see, this triangulation is a crucial part of what makes it possible to keep hold of the real world out there and continue to build a shared picture of reality, even in situations in which we can't get that direct, interactive contact that we feel when we pound our fist on the table (or stub our toe on it!).

So when we use instruments in these ways to find out about, for example, the composition of ordinary sunlight, we typically don't just rely on one result. As we've just glimpsed, there are many different kinds of instruments we can use, and each one can be used in many different ways. We use all these results to triangulate on the phenomenon we're interested in.

REALITY-BASED DECISION-MAKING WHEN WE DON'T GET INTERACTIVE EXPLORATION

What does this tell us about how we approach decisions, especially when they need to be made with other people, or with a whole society? Maybe the most dramatic aspect of this is the recognition that these practical instruments that extend what we can perceive with our senses are clearly helping us identify a common, shared reality out there in the world. After playing with these instruments and trying to understand something about sound, light, or cholera, we don't find ourselves saying things like, "Well, maybe LED lights and sunlight behave this way for you, but some other way for me." We instead tend to compare notes and show one another results from the instruments with the goal of understanding something together, perhaps to identify a key cause of cholera—and, ideally, of being able to use that understanding to effectively act on the world, by, for example, stopping people from drinking contaminated water.

These sorts of direct but instrumented interactions with the world can leave us with some of that same confidence in our rough knowledge of the world that we feel when we avoid crashing into a solid, heavy table as we're crossing the room. The more experience we have with the best interactive examples, the more confidence we can build that it is possible to know important things about the world that we can act on, and that as individuals, groups, and societies we can trust these key fragments of knowledge that make us effective and sometimes powerful in solving problems or building opportunities.

We also have to recognize the cases where we do, currently, struggle with our sense of reality. Let's assume that our closest contact with reality is when we get to manipulate and play with the input, and look at the output. However, there are circumstances in which we have to make decisions in the world—as individuals or groups, or as a society—where we're at a disadvantage because we can't play with the input and look at the output, making our contact with reality more suspect. All sorts of examples come to mind. In medicine, the system you're dealing with—the human body—is incredibly complex, and although doctors can interact with it directly, they don't have the luxury of just trying things out. It's also hard for the doctor to know which of the many, many contributory variables to play with. To make things worse, some of the variables are hard to see, like microscopic bacteria. Maybe foreign policy faces a similar, overly complicated system, and the stakes can be so high that the last thing you want to do is play around. (Perhaps it's fair to say that any system that involves humans is likely to have an ungainly number of variables!)

In fact, these are both examples of another factor that can make it difficult to interactively know that you are dealing with reality: These are situations in which you can't do it over. You can only do it once. As a cosmologist, Saul worries about any phenomenon based on a time scale very different from the ones we are adept at interacting with. We are best when the response time seems immediate, although we might tolerate having to wait to see a response show up on a screen over the course of, oh, two days. As soon as we start talking about things on a time scale of years or decades, we lose the sense of direct interaction—and those are time scales we still live through. When you are dealing with phenomena that take centuries or millennia to unfold, it's hard to get the sense that the data you collect are helping you get in touch with that reality. It's not that interactivity becomes impossible here (we'll talk about those cases in a moment); it's just that the delay between input and output is so large that it's not clear whether the output is even a consequence of our input. This problem obviously affects our decision-making capacity when it comes to the issue of climate change.

Today every society across the globe is making decisions that will affect the trajectory of life on Earth for a very long time. We don't get immediate feedback on the consequences of those decisions. If we lower carbon dioxide emissions, we can't "wait to see what happens," just like we can't wait to see what happens if we don't lower emissions. There is so little sense of interactivity with the system; the output is too far into the future. That's a problem for science as well as for politics and governments. As soon as the time scale gets long enough, we know that lots of other variables have a chance to act too, not just the one that you're manipulating and wanting to see a response to.

For all these examples, it's not that there isn't any reality out there, but that there are many issues for which the reality is very hard for us to establish. That leaves much room for debate. But science doesn't give up when the going gets tough. Instead, people have invented further scientific tools and clever experiments that are all aimed at "triangulating" on reality to help us deal with the situations where interactivity becomes more difficult. And, ideally, they provide a link to a shared understanding of reality in these more complex cases. We can't just go to our corners of the room and pretend that it doesn't matter if two people or groups are acting on conflicting ideas of how the world actually is. Furthermore, if we really are trying to figure out aspects of reality that are hard to get a direct handle on, then we need to proactively find people with a different picture of reality to help us triangulate on what is truly going on in the world, since it is hard to figure out by ourselves how we could be going wrong. As we will discuss in chapter 10, we can also attempt to capture the best of interactive exploration in even more difficult settings by adopting policies that aim for a combination of experimental approaches and iterative improvements.

This story of using triangulation and interactive exploration to build a shared picture of reality is clearly telling us not just about what the world consists of, but also about what is affecting what. In the next chapter, we will specifically focus on this topic of causal relationships, which is so key to our decision-making and planning in the world.

A PLAYFUL EPILOGUE

Here is a fun mental exercise to try with another person, as a way to illustrate the extent to which we already hold some shared sense of reality. Consider each of the concepts on the following list, and ask the other person to do the same. Which of them are you reasonably confident denotes something that is actually out there in the world, and which is just an imaginary element in our internal stories?

An object you see in front of you
An object you see through a pane of glass
An object you see through a magnifying glass
An object you see through a microscope
Germs that cause disease
An object whose existence you infer by comparing the mass and energy measured for particles that entered an experimental apparatus, to the mass and energy of the particles that exited that apparatus
Temporal duration: "a five-hour time interval"
Gravity
The soul
Light itself, as opposed to things that are illuminated by light
Hunger
Romantic love

Beauty
Inflation, in economics
People's preference for the song "Yesterday"
Altruism
Inspiration

We like this mental exercise because it conveys one of the ways that science and philosophy are hard—but fun! (We're assuming that you and your colleague are having as difficult a time getting to the bottom of this list as we do.)

CHAPTER 3

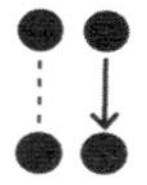

Making Things Happen

Science can transform our ordinary picture of the world. Beyond the features of the world that are easy to observe—mountains and rivers, tables and chairs, people and pets—we now recognize that there is a significant substratum of important things we couldn't previously observe, like biological cells, fundamental physical particles, and the slow but inexorable movement of continental plates. But science isn't just about identifying a list of the elements—and the animals, vegetables, and minerals—that comprise material reality. It is also concerned with figuring out how they interact and influence one another.

Causes and effects become the handles and levers we turn and pull to change our world. If we didn't know about causes and effects, we'd simply be spectators of all that goes on, without any ability to reflect on how to reach in and leverage, manipulate, or change things to make a difference in our own ability to thrive and survive, e.g., the way rocks get sharp when we chip them into flakes, or the way we can make children stop crying by hugging them. But how do we come to know about the handles and levers? How do we come to know about the world not just as observers but as people able to reach in and shape it?

Here we'll look at the problems that the messy real world puts in our way when we try to figure out what causes what, and the techniques science offers to help us work around them. We suspect that many readers are already familiar with the phrase "correlation does not imply causation," perhaps encountered in a high school science or college philosophy class. But a shared understanding of how we actually determine the difference between the two is worth a careful discussion, because that understanding is what enables us to discover the levers for making change in the world. The goal is for all of us to be on the same page on this, so we can first agree on what causes what, and then explain how we got there to others.

CORRELATION AND CAUSATION

Suppose scientists do a survey of people, maybe across a number of countries, to find out the behavioral risk factors for osteoporosis, a condition of weakened bones. And they find that, among the respondents, people who consume more than two alcoholic drinks a day are more likely to have osteoporosis.[8] You read about the results of their survey and wonder uneasily whether this means that *you* have to cut back on your daily alcohol consumption. In fact, the results of the survey don't make that clear. One possibility for the association between alcohol consumption and osteoporosis is that people who have more than two drinks a day tend to also have a sedentary lifestyle, and that's what produces the osteoporosis. In that case, you can keep up the alcohol consumption as long as you also take a few brisk walks every week and maybe do a little light weight lifting. Another possibility is that the alcohol itself causes osteoporosis, in which case, yes, if you don't want osteoporosis, you need to limit your alcohol intake. Well, which is it? And how would we know?

Why is it difficult to find out what causes what? Here's the basic problem. You start out by noticing that two factors are related; for example, that people who consume more alcohol are more likely to have osteoporosis. The trouble is that there are a lot of different causal structures that might explain that relationship. Consider these possibilities:

Model A: No observed relationship

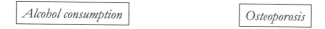

Model B: Alcohol consumption level is a cause of osteoporosis

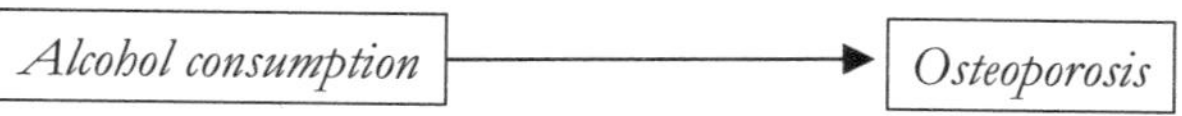

Model C: Osteoporosis is a cause of alcohol consumption level

Model D: Osteoporosis and alcohol consumption level are effects of a sedentary lifestyle

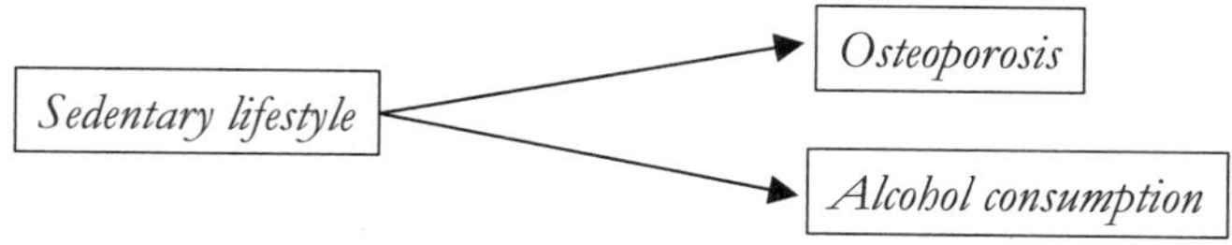

All of these models depict possible causal patterns that might explain the association between alcohol consumption and osteoporosis. In model A, which our data already rules out, these two factors are completely unrelated, or the correlation between them is so small as to be practically negligible. But the remaining three models are all still viable explanations, and we don't know which one is the right one.

Model B is the kind of interpretation many people adopt when they encounter a correlation between a behavior and a disease state. In this model, alcohol is raising your risk of osteoporosis, and the implication is that you should reduce alcohol intake if you want to reduce that risk. Now, there are many reasons you might benefit from scaling back on your alcohol consumption, but if you believe your drinking is otherwise unproblematic, you may want more evidence before scaling back on the basis of this correlation.

Model C also posits that the observed correlation is causal, but in this case, the direction of causality is reversed. In our example, it may seem farfetched that osteoporosis causes you to consume more alcohol, but it is hardly impossible. And indeed, if we replaced "osteoporosis" with "unemployment," model C might seem like a plausible alternative to model B. (And, perhaps, there is a link between unemployment, osteoporosis, and alcohol consumption.)

Finally, model D suggests that the observed correlation is not directly causal, or at least not due to any direct causal influence of osteoporosis on alcohol consumption or alcohol consumption on osteoporosis. Instead, they are correlated because they are each caused by a third variable ("sedentary lifestyle"). Under this model, you might still want to reduce your drinking, but you shouldn't expect it to reduce your osteoporosis risk—it would be more effective to adopt a more physically active lifestyle for better bone health. Note that the "shared cause" factor needn't be "sedentary lifestyle"—there are any number of plausible third factors (food preservatives, radon, pollution, extraterrestrials), and it might take researchers decades to discover the most important one.[9]

So how would you find out which picture accurately depicts the relationship between alcohol and osteoporosis? A great deal of science addresses questions like this. Note that the question presupposes that we've already observed the relevant correlation. To do so, we need good measures of each variable, enabling us to use statistical methods to determine whether they are correlated and establishing that the level of one factor rises with a rise in the other factor, and falls with a decline in the other factor.

For example, for the last two hundred years, bread prices in England have steadily increased, and so have sea levels in Venice.[10] We doubt many doctoral students would stake their career on investigating this correlation as a dissertation topic—it just doesn't seem very plausible. (We'll say more about what "plausibility" might mean later in the chapter.) There are many cases where two factors vary together, and trying to think of any causal network that might be connecting them is an opportunity for comic creativity. Do an internet search for "Spurious Correlations" and you'll find a website containing dozens of examples. For instance, according to the website, per capita cheese consumption has been rising steadily in the US from 2000 to 2009; so, too, exactly in step, has been the number of people dying from becoming entangled in their bedsheets.

Now, it might theoretically be possible that sea levels in Venice cause price increases in British bread, were Venice a major producer of wheat grain for England. But we know it isn't. And it is even harder to think of a good reason why bread prices might raise Venice's water levels. There are two more plausible possibilities. One is like model D; there might be some third cause of both variables—say, some kind of continental weather pattern. But there's another possibility, which is that the apparent correlation we observed in our sample of data emerged solely by chance, and we'd be very unlikely to observe it again if we drew new samples of data. In other words, our sample data misled us, and the correct model was model A above (no relationship).

To try to rule out this possibility, researchers use statistical methods to check whether the observed correlation in this data set is larger than we'd expect to see purely due to random variations in each variable. When scientists say that a relationship is "statistically significant," this is what they have in mind, rather than some claim that the relationship has practical significance for our lives. They are trying to find out whether there is just *no* causal network underpinning the relationship—as in the bread prices/water levels case, or model A above—or if there is *some* causal network that explains the relationship, though it might be the kind we find in any of models B, C, or D.[11]

Okay, so how do we find out whether alcohol consumption is in fact a cause of osteoporosis? How do we cut down the number of possible causal structures that might be explaining the correlation we find?

It's difficult to find out what causes what, but it's essential if we want to reduce undesired outcomes and promote desirable ones in our lives. Consider the case of a virus spreading through our community. It's great to know what is happening and the scientific laws governing what's happening; that this is the level of infection so far, and that in a year, we can expect the level of infection to be ten times that, and so on. But we're not just spectators. We want to know how we can make a difference. Will wearing masks affect infection levels? Will the kind of food we eat influence the virus's impact on us? Does social distancing really do anything to prevent the virus from being transmitted among people? These are all questions about causes and effects. If science is really to help, it has to give us more than a brightly detailed picture of how the world is and what we as observers can expect to happen. It has to tell us about the causes and effects of the virus, and provide information that can help us figure out *what to do*—what the effects of our actions will be in light of those causes and effects. If science didn't tell you about what causes what, then it's hard to see how it would be of any relevance to practical decision-making.

EXPERIMENTATION: THE "GOLD STANDARD" FOR TESTING CAUSATION

Over several centuries, scientists and philosophers have identified a number of criteria for distinguishing among causal models to find the one that can best explain observed correlations. Later in the chapter we will list the major ones, all of which are still in use. But we want to start at the top—the single best method we've identified so far, which is to *conduct an experiment.*

The basic challenge we face is that if all we can observe are correlations, and *causation* is something unobserved, underlying the correlations, then how can we ever know about it? All we ever get are more correlations. This basic puzzle led, historically, to a lot of skepticism on the part of scientists about the very idea of establishing the existence of particular causal connections. But the answer we currently rely on is that a key way to find out about causation is to look at what correlations we observe *under the conditions of an experiment.*

In popular parlance, "experiment" means "Let's try some things and see what happens," but we have something more rigorous in mind here.[12] A scientific experiment is a procedure that has two key features: first, we do "try something" to see what happens—we "intervene" in the causal system; second, we do so in a way that helps us rule out the influence of any possible causes other than the one we are testing.

Before we discuss the logic of intervention, let's look at the scientific recipe for ruling out other causal factors. One ingredient is that we try to *standardize* as many factors as possible—we try to keep them from varying for irrelevant reasons. For example, if we are testing whether an instructional method is effective, we'll want to minimize differences in the way instructions are given across participants or sessions (e.g., by providing a lesson plan and script), and we'll try to use a consistent method of measuring each participant's performance (e.g., by using the same quiz questions).

A second ingredient is a *control group*. In our test of a new instructional method, our control group would be a group of students who don't receive instruction using that new method—getting no instruction at all, or perhaps being instructed using a more traditional method than the one under study. We can observe someone who has been exposed to our instructional method, but to know whether it mattered, we need to infer how they would've done if they hadn't received it—what philosophers and statisticians call the counterfactual.

To make the control group and the "treatment" group as similar as possible, we can try to *match* them as closely as possible. For example, for every sixteen-year-old female we assign to the treatment group, we might recruit another sixteen-year-old female to serve in our control group. Methods for matching have become quite sophisticated, but there are limits to what matching can achieve. It only controls for the variables we used for the matching—in this case, gender and age (but not other differences, for example economic background, nutrition, stress, etc.).

It wasn't until the early twentieth century that scientists (especially the statistician Ronald Fisher) realized that there was a way to control for any irrelevant variable, not just the ones we know. The solution is to *randomly assign* people to the two groups. Statistically, if we flip a coin to decide which of a hundred people are in the treatment group and which are in the control group, then we are likely to end up with roughly the same number of women in each group, the same number of sixteen year-olds in each group, and—here's where things get amazing—any conceivable variable, including ones we would never think of standardizing. There will be roughly the same number of Taylor Swift fans in each group, the same number of Libras, the same number of golf enthusiasts, the same number who came to school hungry, and so on. For this reason, experiments with random assignment are considered the "gold standard."[13]

DON'T JUST STAND THERE; DO SOMETHING!

Ideas from machine learning about the relationship between interventions and causation came as a bombshell to John when he learned about them in 2002. As a philosopher, he had been puzzling over the following problem. Physicists talk very cheerfully about causation: the moon causes the tides, the collision of these two particles caused the emission of a third particle, and so on. So it seems like they take causation to be a pervasive feature of our world, from the particle level to astrophysics. And physics aims to analyze all that exists in the physical world. So just as there are models and systems of equations telling us about mass, electrical charge, quarks, strangeness, and charm, one might expect there to be models and systems of equations telling us about the causal relation. But to John, physicists didn't appear to be treating causal relations this way. Why not? Are physicists being slapdash or even reckless when they describe themselves as having found out what causes what, but don't explicitly show causation in their models and equations? We have systems of equations governing electrical charge. Why aren't there systems of equations governing the networks of causal relations themselves?

The new idea from machine learning was that, as computer scientist Judea Pearl puts it, talk about causation is "a summary of what happens under interventions." Rather than thinking of causation as a mysterious "we know not what" underlying the correlations we observe, we could instead think of causal facts as having to do with what is correlated with what when we intervene in a system. That sense of mystery about causation had haunted the philosophy of science for centuries. But suddenly it seemed evident that we have a simple way of thinking about causation: it's a matter merely of the correlations we observe when we intervene in a system.

Suppose we have a complex system, with lots of interacting parts—the human body, a nation's economy, or a computer circuit, for example—and we want to know how it works. The first diagram below represents such a system, where the "egg" shape depicts the boundaries of that system, and the letters represent the myriad parts of the system. It's a complicated jumble of variables, and we're not sure how any of them are related. Then, as in the second diagram below, we notice there's a correlation between two of the variables, X and Y. Just to make it concrete, suppose we measure a lot of variables that are potentially relevant to health, and we find a correlation between magnesium and heart health: higher levels of magnesium are related to good heart health.

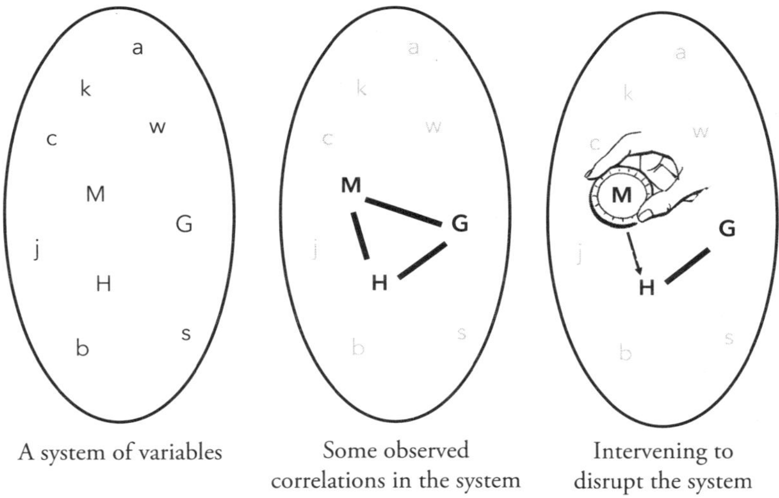

A system of variables

Some observed correlations in the system

Intervening to disrupt the system

Note. M = magnesium, G = genetics, H = heart health

Now what we're wondering is: Does magnesium actually protect against heart disease? Or does something else explain the correlation? For example, maybe the correlation occurs because high magnesium levels and good heart health are both consequences of some genetic predisposition (variable "G").

In the system we are studying, genetics, magnesium levels, and heart health each have lots of different causes, and lots of different consequences, so it's hard to sort out the causal directions. Here's where an intervention helps. When we "intervene in the system," we reach in from outside the system and disrupt it to see what happens. For example, we randomly assign one group of people to take extra magnesium and one group to stick to their regular diet. By doing so, we disrupt the ordinary pattern of factors that cause magnesium levels, e.g., maybe our genes were causing our magnesium levels before, but now our magnesium levels are being caused *by the experimenter.* Thus, note that in the third diagram above, if we now find a correlation between magnesium and health, we can be confident that the direction of causality is from M to H.

We can intervene in any of the relevant variables. If we vary magnesium consumption while holding everything else constant, we can find out whether it has a *causal impact.* On the other hand, maybe good health causes us to get more magnesium. If that were true, then we could intervene by making half of the people healthier in some way (for example, get them to do more exercise), to see whether doing so raises their magnesium levels.

In this way, an *experiment* is when you come from outside the whole system you're interested in and interfere with it, to see what happens. For example, suppose you notice that the position of the needle on a barometer is related to the upcoming weather. And you wonder: Is the position of the needle *causing* the weather? How would you find out? Open the barometer case, reach in and grasp the barometer needle with your hand, and push it to whatever position you like. If the weather changes as you push the needle with your hand, then the position of the needle is indeed a cause of the weather. If not, the position of the needle is not causing the weather. It's just a sign of some other factor, such as air pressure, that is causing the weather—and, presumably, the position of the needle on the barometer.

Another example: If you were looking at a complicated electrical circuit and trying to figure out how it works, you might come from outside the system, and use an electrical probe to fire up a part of the system. If something else fires up when you do that, you can conclude that one is making the other happen.

HILL'S CRITERIA

There are many cases, though, where deliberate experiments are difficult or impossible to do, for various practical, financial, or ethical reasons. For example, almost every large galaxy observed has been found to have an enormous black hole at its center—or, to use the technical term, a "supermassive" black hole. The gravitational mass of such a black hole may be millions or billions of times the mass of our sun. Astrophysicists have observed a correlation between the mass of a central supermassive black hole and the total stellar mass of the host galaxy. Does the mass of the black hole determine the mass of its host galaxy? Or does the mass of the host galaxy cause the mass of the black hole at its center? Or is there some common cause for the mass of both of them? It will be a long time, to put it mildly, before we can manipulate the masses of these black holes, or of these galaxies, to find out what is causing what. Similarly, geologists might want to know what caused what in the steps that led to the formation of a mountain range, but we can't now manipulate what was happening millions of years ago. Can we still find ways to determine what's causing what here?

Or let's look at a more down-to-earth example. The causes of cancer are hard to pin down because (among other reasons) many years can pass between cause and effect. It's hard to know what someone was exposed to in the past that might have caused their cancer. Nowadays researchers identify new environmental causes of cancer so often it can seem like you can't pick up a food ingredient without seeing a label saying it may cause cancer. But finding causes of cancer began somewhere. The first person to spot an environmental cause of cancer was the English surgeon Percival Pott, in 1775. Pott noticed that there was a correlation between exposure to soot, an occupational hazard of chimney sweeps, and cancer of the scrotum, which chimney sweeps got more often than other people. As we've seen, correlation doesn't imply causation, but how could you do a decisive experiment here, to find out what's causing what? A randomized control trial would require you to put together a batch of human subjects whom you'd randomly divide into two groups. Then, arbitrarily, you'd choose one of the groups to expose to lots of soot. The others would remain pristine. And then you'd look to see whether there was any difference between the two groups in scrotal cancer. It's not hard to see ethical reasons this experiment should never have been done, although sadly there are analogous historical precedents.[14]

Pott went about things in a different way. He was looking at a very big correlation between soot exposure and this form of cancer. Basically, so far as Pott could tell, chimney sweeps were about the only people who got scrotal cancer, and they were also the only ones exposed to large amounts of soot every day. (Even as recently as the 1920s, the deaths of chimney sweeps from scrotal cancer were about two hundred times those of workers who were not specifically exposed to the tar or mineral oils in soot.) Did this mean causality? Pott reasoned that the size of the correlation made a difference.

Could some third factor, other than soot exposure, explain this enormous incidence of scrotal cancer in chimney sweeps (for example, suppose that all chimney sweeps ate the same traditional chimney-sweep breakfast)? Pott argued that there simply wasn't anything else, other than soot exposure, that was special to chimney sweeps and might explain this enormous difference in cancer rates.

In 1965, Austin Bradford Hill, a distinguished epidemiologist and statistician, one of the people who established the connection between tobacco and cancer, published a famous paper that addressed the problem of finding causal connections when you can't do decisive experiments.[15] One of Hill's points, drawn from Pott's study, was that if you're getting a *big* correlation between two factors, that improves the case for saying that one is the cause of the other. And that's particularly so in cases like that of the very elevated level of scrotal cancer among chimney sweeps, where it's difficult to see what third factor might be in play.

Another factor we can look at is what Hill called *consistency*: Is the correlation found in many different contexts? For example, you might look at whether you're getting a strong correlation between soot exposure and scrotal cancer in different countries. Does this happen at a similar rate to chimney sweeps in Poland? In Azerbaijan? In Australia? If you're getting consistency across different contexts, then this strengthens the case for a causal relationship. Consistency is getting at the same point as randomization. You're seeing a relationship across lots of different backgrounds, where the other factors that might be relevant to the disease are scrambled.

Another of Hill's criteria is *temporality.* We know that cause precedes effect in time. If you've got two variables, the one that happens first in time can't be the effect. So if the presumed cause occurs before the observed effect, that's consistent with it being the actual cause, and you can rule out the possibility that causality goes in the other direction. It is this criterion of temporality that allowed Pott to rule out the possibility that the strong correlation between soot exposure and scrotal cancer was due to cancer somehow causing soot exposure—or causing people to become chimney sweeps.

Another thing you might look at, in getting from correlation to causes, is not just the size of the correlation, and whether it holds in lots of different contexts, but whether there's a *dose-response* relationship between the two factors. If you look at your chimney sweeps and you find that the more soot they're exposed to, the greater their likelihood of scrotal cancer, then that improves the case for saying the soot exposure causes the cancer. Similarly, if your risk of lung cancer increases with the number of cigarettes you smoke, then that helps make the case for cigarette smoking as a cause of cancer. If one is causing the other, then we would expect a dose-response relationship.

The next criterion we should mention is what Hill called *plausibility.* Do we know of some mechanism whereby A might be affecting B? Or does our current knowledge not really fit with the idea that A might be affecting B? If you can come up with a plausible mechanism for how the presumed cause brings about the effect—if you see a possible way for soot to produce scrotal cancer biologically—then that again improves your case.

We've stuck with this single example of chimney sweeps, but it's obvious that these ideas for getting some insight into what's causing what, without having to do the critical experimental intervention, apply to lots of cases. If you're wondering whether smoking causes cancer, for example, direct experiments would not be doable for ethical reasons, but Hill's ideas will apply. Indeed, these are the ideas that have convinced most people today that smoking does cause cancer. If we find a large correlation between smoking and cancer, and the correlation is reproduced across many different contexts, and there's a dose-response curve (the more cigarettes you smoke, the greater the danger of cancer), and we know a biological mechanism that might explain how one is causing the other, then even without an experiment, we can have high confidence that smoking causes cancer.

These ideas, developed for these down-to-earth cases, can also help us with the astrophysical example we started with. Suppose we were in some way able to determine that many supermassive black holes were born *after* the completed formation of their host galaxies; that would certainly weaken the case for the existence of the supermassive black hole being the cause of the mass of the galaxy.

Hill gave further criteria. So far we've mentioned the *size* of a correlation, the *consistency* with which it's observed across different contexts, the existence of a *dose-response relationship* between the cause variable and the outcome variable, *temporality, and plausibility.* And finally, we can sometimes argue by *analogy* from one area to another. If we've found that exposure to soot causes cancer, and we're wondering if tobacco smoking causes cancer, one line of argument that might support this idea is to find analogies between the way soot acts on the body and the way tobacco smoke acts on the body.

Hill didn't intend these ideas to be anything more than provisional, rough-and-ready criteria that we might use for finding causal connections in the messy real world. Nonetheless, they are now canonical, often cited by researchers in different fields as evidence for causal theories.[16]

SINGULAR AND GENERAL CAUSATION

So far in this chapter, we've been talking about what we might call general causation: things like "smoking causes cancer," or "love causes pain," that are relations between *general* characteristics, *smoking* and *cancer,* for example. That's the kind of thing that randomized controlled trials, and all the other techniques we've been discussing, are meant to establish.

But suppose there's a big oil spill in the Gulf of Mexico. We want to know what caused it. This kind of question is practically very important. If there's compensation to be paid, if someone's to be identified as the one who should clean up the mess, we need to know how the oil spill happened. Suppose we're told that there was one principal cause of the spill, that it had to do with a cracked cement barrier under the water. These are two specific events: the cement barrier failing and the oil spill appearing. One is the cause of the other.

But this isn't a question about general causation. It's not a question about the relationship between two *types* of things. It isn't in general true that cement barrier failures cause oil spills. There are plenty of times cement barriers fail yet there is no spill. To give another, perhaps more tragic example, suppose someone is in a car crash and the seat belt gets wrapped around their neck and chokes them. In this case you would say: the seat belt caused the death. But it wouldn't be right to say that in general, seat belts cause deaths. On the contrary: seat belts save lives.

Now, in these kinds of cases, you know the sort of thing you would have to do to establish causation. You might look at where the oil particles on the surface of the Gulf came from. You look at what route those oil particles took from their source. That's what will tell you the origin of the spill. When you can follow their track, from the cement failure to the surface of the Gulf, then you've established the origin, which implicates the cement crack as the cause of the spill. This needs a scientific inquiry, but it's quite different from studies involving randomized controlled trials. We're asking a different kind of question when we ask for the concrete cause of an individual episode.

Why is this distinction between a singular causation and general causation important? Since they both have to do with causal relationships, it's easy to get confused about which one you are trying to establish for a given purpose. You can easily imagine an argument that says that a company that makes a dangerous product (say, cigarettes) shouldn't be responsible for the consequences because it can't be proven that any particular consequence (say, cancer) was caused by any particular use of the product. This is clearly a correct understanding of the singular causation claim, but it would not be correct concerning the general causation claim. It is interesting to think about how our society structures its response to these different sorts of causal claims. Typically, we expect that regulations will be put in place to prevent irresponsible actions that lead to hurtful outcomes through general causation (e.g., a faulty design that will injure many people who buy a certain light fixture). But we also allow people to be sued (and blamed) for irresponsible behavior that led by singular causation to a hurtful outcome (e.g., an incorrectly installed light fixture that falls on someone's head), and in fact there is a whole range of legal mechanisms that handle both singular and general causation.

GRABBING THE CAUSAL LEVERS—WITH CAUTION

If we can establish what causes what, then we open up the possibility of changing our world—ideally for the better. We can treat diseases, solve famines, educate our children. However, a theme running through this book—*the* running theme?—is that we need to be able to watch out for the ways that we get things wrong. Our understanding of causal connections in complex conditions is almost always imperfect—and we need a way to account for this uncertainty that still allows us to move ahead and take reasonable action when it is appropriate.

This brings us to our next 3MT topic: probabilistic thinking.

Part II

Understanding Uncertainty

CHAPTER 4

A Radical Shift to Probabilistic Thinking

As we begin to discern what we know about reality, two things very quickly become evident: there's a lot we don't know, and much we're still uncertain about. Uncertainty can make us anxious. We're humans, physiologically wired for survival: if we don't know what's lurking in the forest, it makes sense to be wary of venturing in. But the truth is, knowing what we don't know, or only partially know, is also vital to surviving—and thriving. That brings us to perhaps one of the most fundamental currencies of scientific thinking—one we see as central to 3MT: the use of uncertainties in a way that allows us to have strong confidence in what we are doing.

Science offers us a radically different way to think about our connection to this reality that we know something, but not everything, about. It allows us to shift from a stance that says we can only work with things that we are absolutely certain of, to a stance that says we are actually more successful if we can work with things that we have varying degrees of confidence in. Furthermore, just understanding the concept that confidence comes in degrees can be much more powerful than holding out for definitive answers in a world where the available evidence usually fails to give us the absolute certainty we'd like.

Imagine that you are trying to ski down a slope with your knees locked and your stance rigid, without adjusting your weight or bending your legs. It sounds like a recipe for disaster. To stay upright while skiing, you need the stability that you get from constantly shifting your weight from one leg to the other, a *dynamic* stability. Similarly, when we make decisions that need to be based on knowledge about reality, we don't want to rigidly insist that everything we currently believe must be true. Rather, we need to put more weight on some beliefs and less on others, and shift these weights as we learn new facts about the world, so that we can update our decisions as necessary. This is one of the most important but rarely expressed tricks of science, one that gives us mental flexibility to negotiate the "ski slopes" of our uncertainty in understanding the world. We call it probabilistic thinking.

The shift toward probabilistic thinking has been slow in coming, and it is still far from complete. Many people continue to have a brittle, binary view in which any empirical claim—about the efficacy of a new medicine or diet or criminal justice policy—is either right or wrong. From this perspective, any counterexample—"My uncle got the vaccine and still caught the flu"—is seen as a complete discrediting of the original claim, and perhaps a reason for the scientists who made that claim to feel embarrassed.

Scientists have moved away from such black-and-white thinking by building up a culture in which you state any proposition with built-in tentativeness, and that very tentativeness—the practice of attaching a degree of uncertainty to every statement—is a big source of science's strength. It helps you to not feel too attached to any particular belief you have at that moment. By not having your whole sense of self rest on your statements being right every time, you give yourself room to be a proud and confident scientist and yet be wrong some of the time when you say, "I'm fairly confident this theory captures what's going on." (In the skiing metaphor, you're putting your weight on different propositions with different amounts of strength, that is, different probabilities of being right.) In fact, instead of staking your identity on always having to be right (an impossibility), the goal is to be invested in your ability to judge roughly how confident you are in something. Just as a skier learns to look ahead to reduce unwelcome surprises, by acknowledging uncertainty scientists learn to look ahead for reasons they might be wrong. This probabilistic stance is an essential building block of 3MT, and brings many benefits and capabilities. We can think of it as a jiujitsu move to take our weakness (uncertainty) and turn it into our strength.

Scientists learn several good habits for expressing uncertainty. One is, when possible, to quantify our predictions by assigning a number—a probability—to them. For example, if you google "chance of a major earthquake in the Bay Area"—something we've googled more often than we care to admit—you will find statements like "Within the next 30 years the probability is 72 percent that an earthquake measuring magnitude 6.7 will occur in the San Francisco region."[17] In one sense, the statement conveys that we are fairly ignorant about the likelihood of earthquakes, but it also reveals a tremendous amount of knowledge. The statement tells us that scientists must be modeling risk over time (hence "next 30 years") and place ("San Francisco region"), and that oddly specific "6.7" suggests that they probably have some reason for using that as a threshold—perhaps some feature of their theory or the data available to them.

If a scientist wants to say that they are very, very confident something is true—that it's obviously, definitely, absolutely true—their training nevertheless tends to make them reluctant to say it's "yes, 100 percent—it is obviously, definitely, absolutely true." They might instead say they have confidence in it at the 99 percent or even the 99.9999 percent level. Saying that you have confidence in something at the 99.9999 percent level is essentially saying, "I would bet my life on this being true." But it also says, "I acknowledge that I could turn out to be wrong." That ability to step back from an absolute statement is a first key to this probabilistic thinking superpower. (It may be that every swan I've ever seen is white, but "no amount of observations of white swans can allow the inference that all swans are white, but the observation of a single black swan is sufficient to refute that conclusion"—as the 19th-century philosopher John Stuart Mill famously argued, elaborating on a point of David Hume, and apparently paraphrased by Nicholas Taleb.[18])

Of course, there are many things you're not 99.9999 percent confident about. In fact, one of the most productive aspects of science is that we're constantly revising our knowledge and learning and discovering new things, and the world itself is dynamic. We're still constantly taken by surprise by how the world works. We need a way to be able to talk about the world that captures our hard-won awareness that our knowledge about it is a work in progress. We need to be able to say things like, "I really think that this understanding of how the world works is likely to be right; in fact, I give it an 87 percent probability of being right." And we need to be able to express even greater doubt. For example, "I give this new theory just a 51 percent probability of being correct." This range of confidence from 0 to 100 percent is one of the tools of science that we all can use to deal with the world. (Later we will discuss the methods scientists use for calculating confidence.)

It's interesting to think of the development of this probabilistic tool of scientific thinking as the latest in a series of steps that we have taken over the centuries as we develop a more complete battery of descriptions of our world that help us understand it better and work with it more effectively. We began with naming the things in the world, then started classifying them into categories and hierarchies. Next, we were able to measure and otherwise quantify their properties, and now we have begun to quantify our confidence in those quantifications!

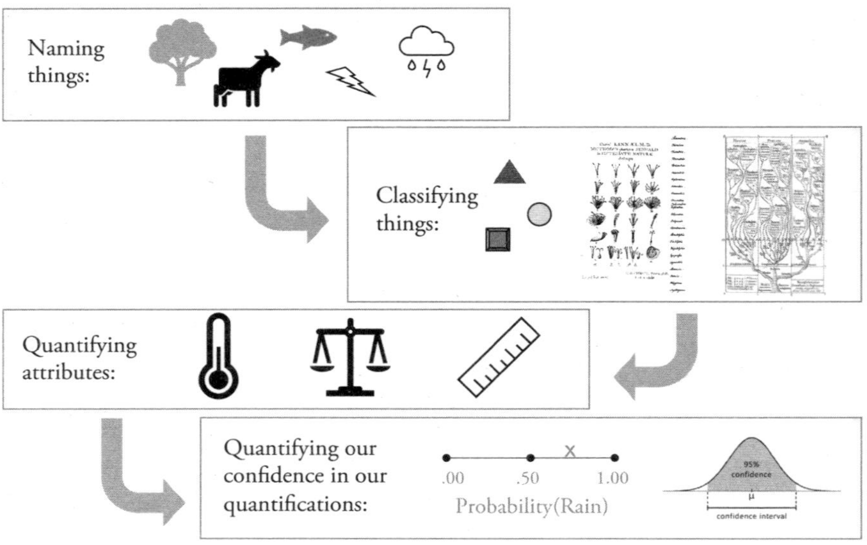

THE STRENGTH OF BEING UNSURE

Why is this tool so important? Perhaps most obviously, it enables us to make fluent use of partial information. For example, suppose you want to build a bridge held together with bolts, but you know that bolts can fail. Simply by learning the *probability* of a given bolt failing over the life of a bridge, you can proceed to build the bridge: Just make sure that every important connection has sufficient redundant bolts that the odds of *all* of them failing on any bridge connection are low enough to bet your life on. (More realistically, you want to bet your life on a *sufficient* number of bolts not failing that the structural integrity is preserved—but here's where you should go find a good engineer to get it right!) If you couldn't use this probabilistic information you would be stuck, you couldn't trust your bridge—since almost no aspect of our real physical world comes with perfect guarantees. Learning these probabilistic techniques of modern engineering opened up a world of actual construction possibilities we wouldn't otherwise have.

A more subtle but powerful advantage of thinking probabilistically is that it gives scientists a way to appropriately save face when they are wrong because it allows them to be wrong without losing their credibility—since as a scientist, almost anything they've said was said with some degree of uncertainty. And scientists aren't the only ones who can benefit from this technique.

It's surprising how important it is to have an appropriate way to save face when you turn out to be wrong about something. The need to save face apparently has roots deep in the earliest stages of childhood. One study[19] of lies told by two-year-olds suggested that one of the biggest motivations for very young children to lie, apparently, is the desire to save face and not show they've made a mistake. In one example, a two-year-old who was asked, "Where's your daddy?" replied, "He's upstairs." On then hearing Daddy outside the back door, the two-year-old said, "My *other* daddy's upstairs." (In this case, there was no other daddy.) What is that two-year-old trying to save face over? It's not like we really hold our two-year-olds to a high standard of credibility in these things, and yet, it's clear that there is a powerful human drive not to be caught being wrong.

It is interesting to contrast this two-year-old's story with another one involving a physics professor Saul used to work with. He was a respected scientist who had once found what appeared to be the magnetic equivalent of a charged particle—a so-called magnetic monopole. Now, this would be a monumental finding if it held up: You can find an electrically charged particle that has a plus charge or a minus charge, but you can't find a magnetically charged particle that has only a north pole or only a south pole. Magnetic particles always seem to come with the two ends, a north and a south, which you have seen when you've played with magnets.

The scientist who had found what appeared to be a magnetic monopole published the finding. He reported the result in the way we are recommending: he presented what he had observed; he presented reasons that you could be worried that the discovered particle might not be a magnetic monopole; he presented relevant odds concerning the particle identification; and he concluded simply that the "...facts strongly favor identification of the particle as a magnetic monopole...." Subsequently, however, it became more and more unlikely that this really was a magnetic monopole, as further and more complete analyses of the experiment showed that the discovered particle, though still a surprisingly mysterious one, was a poor match to a monopole. The follow-up journal article made it quite clear that the scientist and his group had changed their minds, and they no longer believed that they had found a magnetic monopole.[20]

Yet the professor's reputation as a scientist was not hurt, because he had presented his conclusion with this scientific, probabilistic approach: "Here is our data. Here is the probability that we have evidence for a magnetic monopole, given the odds of other things mimicking it in this experiment." Apparently, something else did. But his scientific style of speaking had saved the day because he had never pretended to be 100 percent certain, and he preserved the ability to say that he had been wrong, without losing his credibility.

Not only does probabilistic language give us a way to have useful conversations without constantly calling one another out for being wrong, but it also proactively encourages us to consider more carefully scenarios in which some aspect of the world turns out to be different than we thought.

In order to explain events, scientists often need to refer to circumstances and events that might well be the case in a world like ours, but are not in fact actually occurring. They commonly use the label "counterfactual" to describe these possibilities. The counterfactual can help draw a contrast between what you are testing and what is likely not to be the case. We often *presume* that some event or circumstance is counterfactual, based on our best knowledge, but we don't know for sure. Although the label "presumed counterfactual" may be unfamiliar, nonscientists engage in such reasoning all the time, e.g., "I have to rush home to walk the dog because my spouse is away on a trip; I will presume that the rare scenario in which her flight got in early, so she is now at home, didn't happen!"

If you think in probabilistic terms, you are more likely to consider these presumed counterfactuals seriously: "Well, okay, I'm pretty sure this is right; maybe I'm 90 percent sure this is right. But if I were wrong, then what?" This question is probably not so important to ask when the issue is walking your dog, but in other circumstances it might lead—and has in fact led—to very interesting scientific results. So it's a useful practice to turn these "presumed counterfactuals" into "alternative scenarios," and take seriously the possibility that they correspond to factual reality.

There's an exercise you can do that not only helps you appreciate the importance of quantifying your confidence whenever you make a proposition or put a point on a graph, but also instills this approach into almost anything you do. We want you to have a real conversation that's based on acknowledging your level of confidence in the truth of everything that's said. You can use any topic for which there are a variety of opinions out there in the world. For example, let's say you're discussing with a group of friends whether the increased use of standardized testing in K–12 schools has improved the quality of education or made it worse. Anytime during the discussion, when any participant makes a proposition that could be true or false, that person should stop and say a number from 0 to 100, indicating their percent confidence in the truth of that statement. If somebody forgets to do it, the others should interrupt and get the person to stop for a second and give their confidence number. It's also sometimes interesting to pause to focus on the things you are not so confident about—where you just gave a confidence level in the truth of a statement well below 95 percent—and ask, "If I'm wrong, in what way would I most likely be wrong? And what questions should I ask, to learn more about what I don't know?"

Once you've found some friends who are willing to put up with this, you all might discover that some interesting things come out of the exercise—regardless of the subject being discussed. Students who have tried it report that speakers feel more pressure to have evidence for the statements they make when they accompany them with high confidence values. Also, as we hear other people's estimates of confidence in their statements, we tend to want to update our own estimates. And over the course of the discussion, confidence levels drop from around 90 percent to somewhat lower levels as people become more cautious. Some groups observe that many estimates of confidence end up in the range of 60 to 75 percent once people begin to realize that they aren't so sure about the statements they might have made more dogmatically at the beginning of the conversation. People also find that there's an inverse relationship between the specificity of a statement and the confidence they are willing to express in it: In most situations where we don't have a detailed data set, we can be more confident about the truth of a vague and general statement (since we aren't hemmed in to one version of the truth) than we would be about the truth of a specific, detailed statement.

These observations raise an important question: If people in our society had real discussions this way, would it change the dynamics of those discussions? Would it encourage people to be better listeners? To be more cautious in making assertions? More willing to consider alternative scenarios? Perhaps you should try forcing everybody at the dinner table for the next couple of nights to do this with you and see what happens. Or, if that seems like it might result in eating alone, it might be fun to try just a couple of rounds with people and see whether the feel of the conversation is different, especially if you find yourself in some interesting debate.

A FORM OF ABSOLUTE HONESTY

When you participate in or even watch this sort of discussion, you likely can't help feeling that the use of confidence levels is a way to be frankly honest, openly revealing exactly how strong or weak your understanding is on each point. In the culture of physicists (every subfield of science has its own culture!), it's almost considered *dishonest* not to indicate the range of confidence around any measurement number that you report.

Of course, when scientists participate in the real-life analogues of these confidence-informed discussions the bar is generally higher for the source of the confidence number that they are expected to report: Whenever possible, they certainly prefer to use statistical methods, rather than a gut guess. In fact, in any given experiment, much effort is spent on developing principled ways to calculate and express the range of uncertainty.

So you don't say that you measured the distance of the moon from Earth on a certain night to be 229,733 miles, but rather 229,733 plus or minus (±) 9 miles, which means "I have 68 percent confidence that the actual distance that night was between 229,724 (733 minus 9) miles and 229,742 (733 plus 9) miles." The plus-or-minus range (sometimes called a

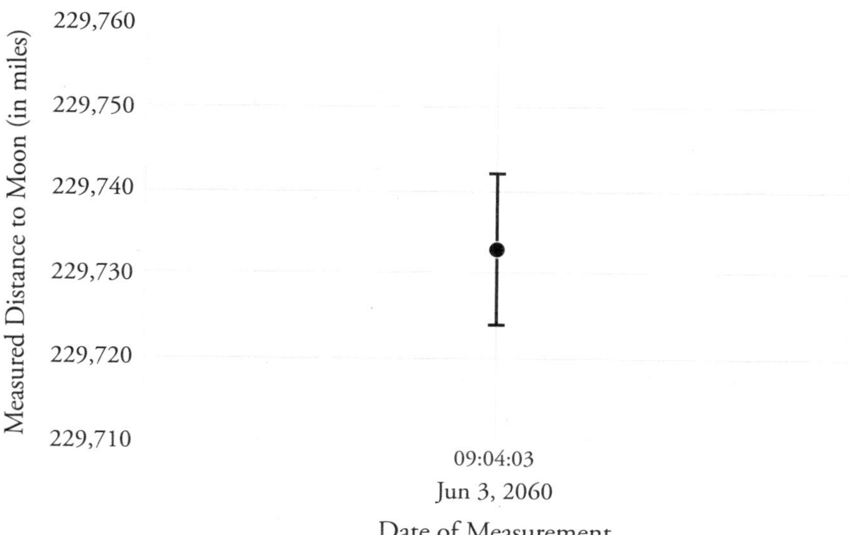

confidence interval) is shown on graphs as an error bar that indicates how much above or below a plotted point the answer might be. For those not used to seeing error bars on graphs, For those not used to seeing error bars on graphs, we've put an example for this moon-distance measurement on the previous page.

Luis Alvarez, the Nobel laureate in physics, used to hold a Monday-night seminar series in his home, up in the hills of Berkeley. Each week, he would invite a physicist to speak on their current research, often a famous professor who might be visiting Berkeley or a prominent member of an international physics experiment collaboration. He would sit in his big armchair and the faculty and students and postdocs and scientists would sit in rows of folding chairs in his living room. He traditionally gave the speakers a hard time, asking challenging questions. Saul still remembers one evening in particular when everybody was sitting in their seats and that week's speaker got up and showed a graph on the overhead projector.

Luis said, "Where do those error bars on your graph come from?"

The speaker said he wasn't completely sure.

Luis said, "Well, if you don't understand your error bars, I don't think there's any point in hearing your talk." And he just stopped the talk right there.

The rest of the group said, "Oh, come on, Luis, let's hear the talk." But he refused.

His point of view—perhaps a rather extreme implementation of that "culture of physicists"—was that if you don't know the range of uncertainty around a measurement, then you have no idea how wrong it could be, so essentially the measurement is not telling you anything. (And perhaps physicists were tougher on one another in those days, even if Luis was an extreme example!)

In a more playful vein, Saul also recollects attending a three-day cosmology workshop, during which the scientists bantered with each other about how to express their confidence levels in various results in cases where they didn't have a good quantitative estimate. The descriptions of confidence levels ranged from, "I'd bet my life on this," and "I'd bet my house on this," to, "I'd bet my gerbil on this," and even, "I'd bet your gerbil on this"!

THE CONSEQUENCES FOR DECISION-MAKING

So far, we've shown you how reflecting on and sharing degrees of confidence in your own statements can make you better able to change your mind when appropriate, and more aware of possibilities that you wouldn't ordinarily focus on. It allows you to put appropriate weight on partial knowledge, engage in more productive discussions, and save your pride and reputation when you need to change your mind or admit you had something wrong. But this quantification of confidence levels can also have further real-world consequences. For example, suppose you are on a jury in a trial where an eyewitness has identified the suspect as the person who committed a burglary. How high would your confidence level in that identification need to be for you to be willing to vote to convict and send the suspect to jail? In the next chapter we'll be discussing better and worse ways to figure out that confidence level, but for now let's assume that you have a pretty good sense of how confident you are (say, on a scale from 0 to 99.999 percent) that the eyewitness correctly identified the suspect. Once you choose a number for how high that confidence level needs to be for you to vote to convict, you start to realize that these numbers have consequences.

Let's look at another real-life example that we experience almost every day: Suppose you're attending a class here at the University of California, Berkeley. Suppose too that you have to cross Hearst Avenue every day to get to it. (Feel free to substitute your workplace and the street you cross to reach it.) What's the likelihood that you will get hit by a car while crossing Hearst on your way to class? Your choices are these: A, about 1 in 1,000; B, about 1 in 100,000; C, about 1 in 10 million; D, about 1 in a billion; and E, about 1 in 100 billion.

It sounds right to a lot of people to put the chances of being hit by a car around the 1-in-100,000 level. But let's look more closely at this number. How often do you cross a street like Hearst? Perhaps a few times a day. That means that almost 1,000 times a year you cross a street where you could be hit by a car. Imagine that you do that every day of your life, on average, and of course we all assume we're going to live for 100 years, right? So if you risk being hit by a car 1,000 times a year, that means in 100 years you will take that risk 100,000 times. If the odds of getting hit are 1 time in 100,000 times, then the odds are starting to look pretty bad for you. You're actually likely to get hit crossing the street!

Now, most of us are hoping that when we die it will be due to other causes than being hit by a car on Hearst heading to class. So most of us should insist on finding a crosswalk where the odds of making it safely across a street are better than 1 in 10 million, if we are going to cross it several times a day. (This means that we have better than a 99 percent chance of making it through a whole lifetime without getting hit.) The point here is that these quantitative estimates of your confidence level can make a big difference. You can make decisions based on these numbers.

POLITICS AND CERTAINTY

Is it always a good idea to assess and share our degree of confidence about the things we say? A problem that immediately comes to mind is: What happens when someone speaks in terms of degrees of confidence, in a political environment? Imagine that in a major speech the president puts forward a new policy proposal on health care reform. Which possible statement would make you feel more confident about it?

A. The policy that I'm putting forward is the right policy for America. I guarantee that it is what's best for the country.
B. I think that the policy I'm putting forward is the one that is most likely to be the right policy for America. There is no guarantee that it will work; in fact, I give it only a 75 percent chance, but the alternatives that have been presented are all much less likely to succeed.

We rarely hear language from politicians like that in statement B. Our guess is that all the political consultants will tell you that you cannot say something like B and expect to get many votes. The reason they're likely to cite is that we are looking for our president to be like the omnipotent parent most of us last had when we were two-year-old children. All of us probably remember that time when we had the vague subconscious feeling that our parents always knew the answer and could solve all our problems. It was amazing how much they knew compared to what we knew at that time. Many of us, as adults, would like to re-create that feeling of security by having a real expert, someone you could always trust, in a leadership role like the presidency.

On the other hand, for us authors, and probably for others reading this book, a statement like B would be so refreshing to hear, right? We've heard statements like A many, many times. And we've had the experience that just because somebody says they know the answer and it's got to be right doesn't mean it is right. A statement like B would give us more of a sense that the person was actually thinking about the complexity of the problem and understood the odds of getting it right the first time. For our money, we would much prefer to see a world in which all the political arguments were couched in language like statement B, because it shows there is room for learning and adapting. But it is obviously not the thing that makes the average person say, "Oh, now that's the candidate I'm voting for." At least not yet. Once everybody has read this book over the upcoming years, we're sure we'll have a different response from the informed citizenry!

By this point, it should be clear why probabilistic thinking is so important to a 3MT citizen. We can change our minds, or have one of our previous claims turn out to be wrong, and keep our self-respect. We can consider alternative scenarios, and not be stuck with one rigid expectation of reality. We can take pride in being brutally honest about what we know and don't know—and, even better, when we are in between, we can honestly quantify how likely a statement is to be true. We can be less defensive and more open-minded in discussions with others. We can productively use knowledge that is uncertain, and proactively hunt for information that might lead us to revise our predispositions. And we can even make effective decisions—to convict or free a defendant, or simply to cross the street—based on these likelihood numbers.

But these key advantages are only the beginning. Probabilistic thinking might be the Swiss Army Knife of 3MT. It is a tool with a thousand possible uses that enhances your odds of surviving and thriving in any terrain. In the previous chapters we discussed how we seek to get better and better understandings of what is really going on out there in the world, what it is made of, and what the causal relationships are that allow us to act on the world to change it, ideally for the better. The probabilistic stance reflects the relationship between our imperfect human understanding and the actual shared reality out there. In the end, all of our randomized controlled trials of a causal relationship (or our Hill's criteria assessments, for that matter) simply provide us with quantitative estimates of how likely it is that a given causal relationship correctly describes our world. And when we have many "logs" of such probabilistic knowledge tied together into the "raft" of our scientific understanding of reality, they support one another and can strengthen our confidence in some tightly bound part of that raft. Or conversely, a section of the raft made from many weakly confident logs can draw our suspicion that something may be wrong with our understanding of that part of reality.

We have been emphasizing our human uncertainties about how the world is constructed and its causal relationships, but it is interesting to add that there are also uncertainties that appear to be built into reality itself. Even if we perfectly understood how some system, say a nearby star, worked, there would still be a true coin-flip-like randomness involved in its behavior; for example, whether it would explode as a supernova in the upcoming decade. Or, more worryingly, we can't know when the next major solar storm will knock out our power grids: Will it be in the next decade? Or the next century? (There was a huge solar storm in 1859, and in the past century several smaller storms that still knocked out power grids, including those in 1921 and 1989.) Similarly, there are situations when we don't know whether a medical treatment will work in a particular case, but we have strong confidence (98 percent) that it has a probability of 70 percent of working. In these situations where the world itself has random aspects, we can use probabilistic thinking to identify, rank, and manage the risks.

We live in a world with innumerable unanswered questions, and there is no back-of-the-book answer key that is guaranteed to tell us how everything works. Being able to build on these probabilistic understandings of our confidence levels is crucial, a source of our strength and effectiveness. It's therefore particularly important to learn how, as individuals, we can steadily improve our use of these probabilistic tools by better calibrating our estimates of confidence levels. And, equally important, we need to understand how the experts we consult use these tools, how well calibrated *they* are when they estimate their own confidence levels, and how well focused they are on becoming better calibrated in their confidence estimates over time.

Developing skill in calibrating our own confidence levels, and an understanding of the skill (or lack thereof) with which experts calibrate theirs, is at the heart of probabilistic thinking, and is the concern of our next chapter.

CHAPTER 5

Overconfidence and Humility

In the previous chapter, we argued that probabilistic thinking was a major conceptual advance in science, and we described confidence levels as one example of such thinking. Occasionally, however, we see even highly competent scientists failing to acknowledge uncertainty. In July 2020, a highly respected scientist announced on Twitter that "US COVID19 will be done in 4 weeks with a total reported death [*sic*] below 170,000." With the benefit of hindsight, we can say that this prediction was wrong—rather spectacularly so, given that as of this writing, US Covid-19 is definitely still with us and has taken more than a million lives. With hindsight, it is usually possible to look back and find some experts who were wrong. But we mention this example not because this expert turned out to be wrong; at the time, he had plausible reasons for thinking the pandemic might play out differently. Rather, we mention it because of his complete lack of any confidence statements ("I am 80 percent confident that..."). He gave not even a hint that his theory of the pandemic might be incomplete or in error.

It seems worth noting that in this example, the expert was offering an opinion on a topic outside his main area of specialization, and that he was offering his opinion on Twitter. We suspect—with at least 75 percent confidence—that if he had been writing for other specialists on the topic of infectious disease or public health in a specialized journal, he would have been much more cautious in stating his opinion—because if he wasn't, the journal's editor and reviewers would have demanded that he restate the opinion with a defensible confidence level or else retract it. Now, it is tempting to give this scientist a pass and say, "Well, he was doing this on his free time, and everyone knows people can say whatever they want on Twitter." But then ask yourself: How many people in the world would have seen his opinion in that specialized academic journal? Surely just a tiny fraction of the number of people who saw (and retweeted) his opinion on Twitter? We would all do well to reflect on another famous quote from philosopher David Hume: "A wise man... proportions his belief to the evidence."

Expert overconfidence can have grave consequences. After the *Challenger* space shuttle exploded in 1986, an investigation found that NASA had officially predicted that there would be one failure in 100,000 launches—about the same odds we considered for getting hit by a car crossing Hearst Ave in the last chapter. Yet other evidence revealed that NASA had strong evidence against this rosy projection. NASA experts had, only five years prior, noted in a report that the historical failure rate for solid-fuel rockets—which were used to boost the *Challenger* into orbit—was one in every 57 firings. Since the space shuttle launches used two such rockets every time, a failure could be expected once in every 28 or 29 launches, assuming the historical failure rate continued unchanged. *Challenger* was the twenty-fifth shuttle launch, so it was almost right on schedule to fail. So something happened within the organization to transform a pretty pessimistic risk estimate into a much more optimistic—and, apparently, unrealistic—one.[21]

In the mid-twentieth century, Lev Landau, a theoretical physicist, offered a pithy description of expert overconfidence among scientists: "Cosmologists are often in error, but never in doubt." This is perhaps an overstatement; after all, scientists do sometimes retract their findings.[22] For example, in 2010, twenty-three experts published an open letter to the Federal Reserve's then-chairman, Ben Bernanke, arguing that his quantitative easing policies would produce "currency debasement and inflation." In 2014, when it was clear that this prediction was wrong, two journalists contacted the twenty-three signatories. Fourteen refused to comment, but those who did said their views hadn't changed.[23] Having earlier mocked these experts, Paul Krugman, a *New York Times* columnist (and Nobel Prize–winning economist), was more open in early 2022 in acknowledging his own error in dismissing predictions that President Biden's 2021 stimulus package would produce high inflation. "I don't want to be like those guys. So I'm currently spending a fair bit of time trying to understand why my relaxed view of inflation early last year has been refuted by events." However, he did go on to argue that his original analysis was correct in its fundamentals, and that the Covid pandemic was responsible for upending the usual patterns in the economy.[24] (Economics is a tough game to play, and even to write about; while Krugman underestimated inflation, it is still not clear who is right about the specific role played by the stimulus package.)

THE IMPORTANCE OF INTELLECTUAL HUMILITY

With these examples in mind, a challenge for expert authority in the Third Millennium is to cultivate what is often called intellectual (or epistemic) humility. Psychologist Mark Leary has been studying this trait for many years, finding that people high in intellectual humility are "more attentive to the strength of evidence regarding factual claims" and "more interested in understanding the reasons that people disagree with them."[25] He notes that "[c]ultures vary in the degree to which they value openness and flexibility and tolerate uncertainty and ambiguity."

At its best, Silicon Valley is a culture that has fostered an openness about errors, as exemplified by the popular slogan "Fail fast, fail often." Of course, the saying is not a celebration of failure for its own sake, but a claim that failures are an inevitable by-product of cutting-edge technological entrepreneurship. There's a similar notion among many scientists who contend that every graduate student is guaranteed to make certain experimental errors, and so the best course is to get them over with by gaining a lot of research experience as early as possible.

Recently, a community of younger scholars in psychology has begun to promote a culture in which researchers admit their mistakes. In their "Loss-of-Confidence" project, the scholars documented findings they had reported but now doubt. And they reported an anonymous survey of 315 scientists that found that 44 percent reported questioning at least one of their published findings. In the majority of these cases, researchers had not publicly acknowledged their loss of confidence, or did so but in a forum other than the journal that published their study.[26]

CALIBRATING OUR CONFIDENCE LEVELS

As we argued earlier, scientific evidence can provide probabilities but not absolute certainties. This implies that it is both foolish and unfair to expect our experts to be *infallible.* Even if they are doing their job perfectly, they will get things wrong some of the time. But it is quite sensible to expect our experts to be *calibrated.*

What do we mean by "calibration"? If the expert is offering the probability of an event, we can look across many different situations and see if the prediction matches the event's frequency. If the expert is making a categorical assertion—"It is a brain tumor"—we can ask them to quantify the probability that they are correct. And if they are estimating a quantity, we can ask them to describe the range of low-to-high estimates that they are 95 percent confident will contain the correct value.

A person is well calibrated when their stated confidence level at the moment of prediction matches their accuracy rate once we find out about the true outcome. We demonstrate what this means for our students by giving them dichotomous questions like, "Which is longer—the Panama Canal or the Suez Canal?"[27] Now, most people haven't looked up and memorized the answer to this question. But we aren't studying our students' mastery of "useless trivia." Instead, we want to know how they rate their confidence in their response to each item. When their expressed confidence level corresponds to the odds of being correct, they are *perfectly calibrated.* For example, across all the times you give a confidence level of 50 percent, you should be right half the time and wrong half the time. Across all the times you say you are 100 percent confident, you should always be right. If your accuracy rate falls below your expressed confidence, then you are overconfident. You are underestimating your ignorance.

In the figure below, we show the results from many years of giving students the calibration exercise. When students report their confidence level as 50 percent, which essentially means they are guessing, they get things right a little more than 50 percent of the time—perhaps because they know more than they realize. But as people express increasingly high confidence in their answers, they are consistently less accurate than they believed they would be. This "classic" calibration pattern, showing a clear tendency toward overconfidence, has been derived again and again in many different studies for many different populations.[28]

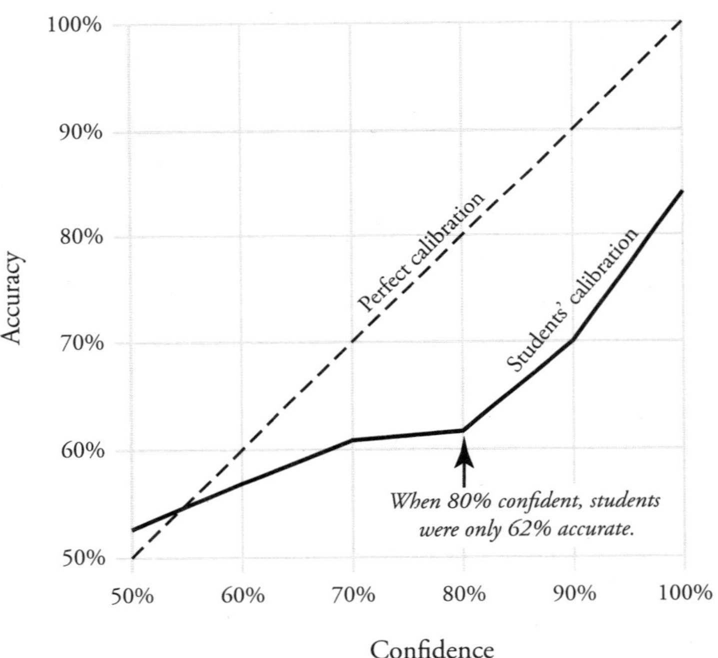

Evidence for this "overconfident" miscalibration is also seen in expert judgments. In the early 2000s, several researchers studied overconfidence among stock market forecasters in Germany. They asked 350 financial experts to give their predictions for the level of the DAX index (the German equivalent of the Dow Jones average) six months ahead of time on a rolling monthly basis. Importantly, they asked each expert to specify a 90 percent confidence interval for each prediction—the range of values within which they thought the actual DAX value would fall nine times out of ten.[29] Here's what happened: Every month, the actual DAX value fell completely outside the confidence intervals that many of the experts had provided six months earlier. In fact, more than half the time during the twenty-six-month-long study, it turned out that fewer than half of the experts had provided confidence intervals broad enough to encompass the DAX value that month. Not only were many of these experts quite wrong about the future direction of the German stock market, but they were also very poor at estimating how wrong they might be.

That previous sentence actually holds the key to the concept of calibration. It contains the idea that in addition to knowledge (what you use, for example, to predict the DAX index six months out), there is *meta*knowledge, or knowledge about your knowledge. The German financial experts who provided confidence intervals that proved too tight were demonstrating poor metaknowledge. They didn't know how much they didn't know. They could have done better by improving their metaknowledge—that is, by calibrating their confidence levels.[30]

Another example comes from Phil Tetlock's research on professional foreign policy experts. What these experts predict can have a profound influence on public policy. Partly on the basis of expert predictions and forecasts, the US Congress makes allocations to the military budget, and the president develops diplomatic, economic, and military strategies and negotiates treaties. The more confident the experts are in their predictions, the more likely members of Congress and the president are to be swayed one way or the other. Tetlock's research suggests that we need to be wary of such predictions. He asked several hundred foreign policy experts to make yes-or-no predictions for events five and ten years out. He asked, for example, "Will Vladimir Putin still be president of Russia in 2016?" For each prediction, he also had his subjects provide a level of confidence on a scale from 1 to 9. He found two pieces of bad news. First, the predictions were barely more accurate than would have been obtained by tossing a coin. Second, there was essentially no relationship between accuracy and expressed confidence. Predictions that turned out to be right had an average self-reported confidence between 6.5 and 7.6; those that were wrong had an average confidence level between 6.3 and 7.1. These were not significantly different. The experts who were wrong were just as confident as the ones who were right. That means the confidence with which a foreign policy expert stated their prediction was a very poor guide for deciding whether you should believe them.[31]

We might expect physicists and other natural scientists to be better at calibrating their confidence than social scientists, especially when they are investigating features of the natural world that have nothing to do with politics. After all, natural scientists have at their disposal data, with frequency distributions and multiple measures, and advanced statistical formulas into which they input reams of data and get back precise confidence intervals. But it appears that those in the "hard" sciences have often in the past had just as much trouble judging the appropriate level of confidence to attach to their findings as financial and foreign policy experts.

Interestingly, part of the reason that we know this about natural scientists is that physicists have been particularly interested in understanding how well calibrated their confidence statements are, so they have been studying and tracking this question for decades. Physics was one of the first fields of science to work with extremely large datasets, and physicists have a long tradition of collaboration amid competition among teams around the world, so in the late 1950s and early 1960s they started to collect, compare, and combine their competing measurements—and their confidence estimates. They soon discovered indications of misplaced confidence in results. For example, in trying to pin down the exact values of physical constants like the speed of light and the mass of an electron, physicists would be expected to report great uncertainty in the initial measurements, followed by progressively more certain estimates over time. In other words, the error bars should start out very broad and get tighter and tighter with each new study, and each new measurement of the constant should be likely to lie within the margin of error of the previous one. But this is not what happened. When the physicists plotted their historical estimates for the value of *c*, the speed of light, from 1870 to the 1960s along with their error bars, they found that the estimates bounce all over the place, and frequently the value estimated in one study lies entirely outside the margin of error given by the previous study. The same inconsistent and seemingly incoherent pattern occurs for the estimates of such physical constants as the inverse fine-structure constant, Planck's constant, the charge of an electron, the mass of an electron, and Avogadro's number.

Of course, during this whole history of the measurements, the scientists believed that *their* work finally represented a close approximation of the truth. For example, in 1941, the physicist Raymond Birge wrote, "After a long and, at times, hectic history, the value for *c* has at last settled down into a fairly satisfactory 'steady' state."[32] Not long afterward, most estimates for the value of *c* began coming in much higher than Birge's estimate, quite a bit outside his stated confidence interval, and the present-day estimate, known to a high degree of confidence, is similarly well outside Birge's "steady state."[33]

After the physicists saw these failures in estimating confidence levels, they became much more cautious about trusting simple internal estimates and began demanding much more cross-comparison of results to gauge the uncertainties, and much tougher standards for accepting a claimed scientific discovery. Yet even so, one of the biggest lessons that physics experimentalists teach their students is that they will *still* be overconfident in their measurements!

Even if overconfidence is a feature of human psychology, improving our calibration is possible. Under certain circumstances, we can calibrate our confidence quite well. When you study the calibration of confidence levels among different professions in which predictions are integral to the work, you find that meteorologists' short-term forecasts, for example, are remarkably well calibrated. If you look at all the times a weather forecaster says the chances of rain the next day are 80 percent, you find that about 80 percent of the time the day turned out to be rainy. Why such good calibration? The key might be that meteorologists are constantly getting immediate feedback about their predictions. In addition, meteorologists' professional prestige depends on their metaknowledge (that is, being calibrated) at least as much as it depends on their knowledge (being accurate).[34]

In any profession or domain, professional demands and social and cultural forces influence people's judgments about the state of their knowledge. Becoming aware of the forces that affect how confidence levels are calibrated in yours may help you identify and resist the forces that subtly push you toward overconfidence. We should strive, in a sense, to emulate IBM's supercomputer, Watson. Watson famously defeated the best human *Jeopardy!* players not just because of its vast, Wikipedia-like knowledge, but because of its astute metaknowledge.

In *Jeopardy!*, metaknowledge is hugely important because only one contestant gets the chance to provide the right "question" to each "answer," and that's the contestant who presses the buzzer first. Since incorrect responses are penalized, there's a great disincentive to simply "buzz in" as rapidly as possible. You want to press the buzzer only if you know, or think you know, the correct response. The contestants who win are those able to rapidly determine if they know the correct "question" or not. Watson is programmed to do this in real time and to do it very well. It knows its own state of ignorance. Watson basically tells you, "In this case you should believe me; in this other case there's little reason to believe me," and that's a very valuable thing in an expert.[35]

CONFIDENCE IN THE CONFIDENCE OF OTHERS

Now, to fully understand expert overconfidence, we have to look at how experts' predictions and assessments are used in the world. We need to see things from the perspective of the "observer"—the patient of the doctor making an assessment about the risk of surgery, the juror considering the testimony of a witness, the investor acting on the stock market predictions of her financial advisor. When we look at cases like these and investigate what kinds of cues people use to decide whether to believe the expert or not, the expert's stated confidence emerges as one of the most important. Advisors, witnesses, and experts are perceived as more credible when they are confident.

One kind of situation where this sort of dynamic comes into play is the criminal justice context, in which jurors hear the testimony of eyewitnesses and need to judge their credibility. Here, the eyewitness is the "expert" and the juror is the "observer." Psychologists researching the question of what cues observers use to judge credibility can set up simulations in which crimes are acted out in public and then actual eyewitnesses to the "crime" are recruited to give testimony in front of mock jurists. They find that perceptions of credibility have a fairly strong positive correlation with perceptions of the confidence of the witness, suggesting that jurors may rely very heavily on the confidence of the witness to decide, "Should I believe this person or not?"[36]

But there's a problem here. We know that stated confidence in a prediction or assessment is an unreliable guide to its accuracy. If people are using an expert's confidence—either perceived or stated—to judge credibility, then they are frequently being misled and may be making poor decisions as a result. Jurors may be sending innocent people to prison, investors picking the wrong stocks, patients electing surgeries that result in complications—all because they are focusing on a person's confidence as a cue for judging credibility and accuracy.

Thankfully, there is evidence that this dynamic can be disrupted. When very confident forecasters, experts, or witnesses are shown to be wrong in psychological studies, their expressions of confidence no longer carry their former weight among observers. The moment the confident person is shown to have made a mistake, people feel betrayed. (Conversely, experts or witnesses who are shown to be wrong but expressed lower confidence in their prediction or assessment don't experience a reduction in credibility.[37])

It makes basic intuitive sense that observers would alter the way they respond to the confidence cue when presented with information about the expert's underlying accuracy. Confidence no longer needs to serve as a proxy for accuracy when the accuracy is known. But the requirement that the observer receive feedback is a critical hurdle. Getting feedback about an expert's accuracy is not always possible. Further, some studies suggest that if obtaining the feedback is effortful in any way, many people take the route of laziness and fall back on the use of the expert's confidence as an indicator of accuracy.[38]

Is it possible for an expert to have it both ways—to avoid overconfidence while also avoiding errors? One way to do so is to make one's confidence interval so wide that it almost surely includes the truth: "I'm 95 percent confident that President Biden will get between 30 and 70 percent of the vote if he runs for reelection." The expert's dilemma is that no one will consider her an expert if she adopts this risk-free stance. (As the range narrows, confidence is necessarily lowered: the same expert might have 70 percent confidence that Biden would get between 40 and 60 percent of the vote, or 60 percent confidence that the range will be between 40 and 50 percent.) Experts need to be calibrated enough to be trustworthy, but specific enough to be informative—a tall order. The good news is that an honest and realistic assessment of one's confidence can preserve trust in your expertise.

CHECK YOUR OVERCONFIDENCE

If experts came in only two flavors—the "accurate ones" and the "inaccurate ones"—most of us would rather hear from the first group than the second. But aside from easy issues—the kind we don't really need an expert for—it is unrealistic to expect any expert to be able to provide insights that are guaranteed to be correct. So statements of degrees of confidence are a crucial source of information for vetting our experts—though not in the way that most readers might be accustomed to!

The next time you listen to an expert talking on your favorite news show, listen carefully to the words they use to describe their confidence. Do they express absolute certainty? Do they use linguistic "hedge" words and phrases like "it is possible that," "a risk of," "one view is that," and so on? In a world of inherent uncertainty, we should prize experts who are *calibrated*. Unfortunately, experts are often under pressure—from journalists, policymakers, attorneys, and the general public—to sound confident.

The famous psychologist Daniel Kahneman once remarked that overconfidence is the one human bias "he would most like to eliminate if he had a magic wand."[39] We doubt that overconfidence can ever be eliminated, but we have seen that there are concrete steps we can all take to reduce it.

We haven't highlighted this, but perhaps the first point is: Don't feel you have to offer an opinion about something you don't know a lot about. Maybe put yourself on a tight "opinion budget": "I will allow myself to state only five opinions today, so I'd better choose them wisely."

If you feel compelled to offer an opinion, state it as a probability if possible, or offer some indication of your level of confidence: "I'm 75 percent certain that..." or maybe, "I think it's more likely than not."

When you listen to experts, pay careful attention to whether they acknowledge their uncertainty and the ways they might turn out to be wrong. We'd like our experts to be 100 percent correct, but that's just not going to happen. But we can and should seek out experts who are close to 100 percent calibrated. Experts who tell you "I don't know enough to offer a firm opinion" are not useless—they are reliable and trustworthy. And if you think they are among the most knowledgeable people on the topic, they've just taught you that this is a topic on which we need more research. In the meantime, if you need to act, you'll know that it's best to act cautiously, and with humility about how much remains unknown.

CHAPTER 6

Finding Signal in Noise

Experienced scientists know that their first understanding of a phenomenon will often turn out to be wrong. And they often learn this the hard way. Saul learned this lesson as a young postdoctoral student when he was working with a number of senior scientists on an exciting astrophysics search that began when a relatively nearby star exploded. Supernova 1987A, as it was called, was by far the nearest supernova—hence the brightest—that had been seen in hundreds and hundreds of years. Scientists all over the world headed to the telescopes in the southern hemisphere (where it could be seen) to study it in all different ways. The team Saul was working with had this thought: When this kind of supernova explodes, it leaves behind an extremely compressed, dense remnant of the star it once was, and this remnant is called a neutron star. That neutron star remnant rotates around and around, and typically it will carry around and around with it a very strong magnetic field, with beams of light (and radio) coming out of its two magnetic poles. If the magnetic pole is misaligned from the rotation axis, the beam of light from one of the magnetic poles can point at you once every rotation, like a lighthouse beam. Typically, the rotation happens something like 1,000 times per second, so you can see a regular pulse of light (or radio) every millisecond—or you can if you have a big telescope and a good light-recording device!

The first time anybody saw one of these things—it's called a pulsar—they wrote down "LGM," for "little green men," on the original graph they were making (of radio signals, not optical signals), because they thought that this regular repeated pulse might be a signal of extraterrestrial intelligence. Then Jocelyn Bell Burnell and Antony Hewish, the astrophysicists who found the signal, started discovering a bunch of pulsars in different places in the sky, some of which seemed to be associated with these kinds of supernova remnants. They started to realize that a pulsar was not *quite* as exciting as a signal from extraterrestrial intelligence. But it was still pretty neat. These pulsars are amazing in themselves.

Saul's team thought they might have a chance, for the very first time, to catch a pulsar in the act of being formed right after this explosion. So they built a new instrument, an infrared-photon detector, and flew down to Chile to put the instrument on the telescope.

Sure enough, they detected a signal at just around the expected range of repetition rate, 2,000 times a second. If this signal were a sound—a pulsation of air pressure, rather than a pulsating light—it would be high-pitched, approximately a B note almost three octaves above middle C. (It's useful here to think of the comparable sound of this rapid a pulsation, since we have some intuition about what it feels like for such a sound's pitch to be high or low, but we don't have an intuition for flickering light signals that are this rapid; our eyes don't notice flickering of light when that flicker is happening faster than about 50 times a second.) Oddly enough, this "pitch" of the flickering supernova light was wandering, first getting a little lower (that is, flickering a little more slowly) over a few hours and then returning to its starting pitch over the next few hours. Why was this?

The team thought a bit and then realized there could be a good reason to expect the signal to wander just a little bit, because there's going to be a bit of a Doppler effect happening—that's the change of pitch you hear when a car horn is coming toward you and then moving away from you. Because the earth is rotating, sometimes the instrument at the observatory in Chile was rushing toward the signal, making it "sound" a little higher-pitched. And sometimes the instrument was rotating away from the signal, which should make it "sound" a little lower-pitched.

And for that matter, the earth is also revolving around the sun, and that motion too should vary the pitch. So to ascertain whether and why the signal was varying, the team had to correct for both those pitch-distorting effects. Once they corrected for them, the signal's pitch still varied, but now its variation turned into something that looked quite regular—almost a perfect sine wave. This was exciting, because a sine wave is what you'd expect from the Doppler shift if that pulsar *had a planet orbiting around it.* The planet would cause the pitch to shift in a regular way, backward and forward and backward and forward, with respect to us. It would look just like the sine-wave signal the team was seeing.

At the time of this exciting result, scientists hadn't yet found any planets outside our solar system. The pulsar signal contained the first-ever evidence for the discovery of such an "exoplanet." That's the kind of discovery that makes it fun to be a scientist. You are just playing, and you look, and suddenly there's a signal. It's hard to pick out. You clean up one of the distortions that you know about (e.g., the effect of the earth's motion) and it turns into something that looks exactly like something you hadn't even been looking for—in this case, a planet orbiting around this distant pulsar. The paper detailing the team's discovery was immediately sent off to the international scientific journal *Nature.*

Then, the plot thickened....

Having just read the last two chapters, you might find yourself wondering what the team's confidence level was in this discovery. We have, after all, been emphasizing how central probabilistic thinking is to decision-making in a world full of uncertainties, and thus how important it is to calibrate our own level of confidence in the points we are making, or the confidence level accompanying the statements made by the experts we are listening to. Let's dive a little deeper into how we play with, quantify, and make use of probabilistic thinking, since it's so helpful to be fluent with this tool.

An explanation of a bit of the scientist's jargon is a good place to begin. Much probabilistic discussion involves hunting for "signal in noise," but what we mean by the words "signal" and "noise" is different from what the words usually mean in other contexts.

What do we mean by signal? In communications, it can be any element that a person uses to try to get a message across to another person—like spoken words, letters, music, or a beam coming from a lighthouse. These signals contain the ideas or directions or emotions that you're trying to communicate. But a signal can also be something a bit more abstract. When you're trying to detect something—a shape or a sound or a scent—that tells you something about the way the world is, the signal is the trace or evidence of that thing.

If these two notions point to what we mean by signal, what's noise? By noise, we mean almost anything that will get in the way of detecting the signal. Some kinds of noise can be very similar to the signal, but not actually intended to communicate with you, or not actually providing you the information that you seek about the world. Beginning with some auditory examples (since the terminology originated with radio engineering), imagine you're trying to listen to someone's words and understand what they are saying, but somebody else is talking across the room in a separate conversation. Presumably, the other person's words are intended to communicate with *somebody,* but from your point of view they are "noise" interfering with the "signal" you are trying to receive. Our ability to focus on our conversational partner amid such distractions is called the cocktail party effect.

Noise can also consist of random distractions, like random sounds, shapes, or scents that happen to exist along with what you are focusing on. For example, if you are trying to listen to somebody talking while walking along the beach and the ocean is roaring, the ocean's roar is not intentional in the way that other people's conversations at a cocktail party are intentional. But the sound of the waves is random noise that can obscure your signal.

While it's easy to think of examples of noise in the form of sound, it's important to be clear that noise in the way we are defining it comes in other forms as well, and exists when we try to detect almost anything, sensory or otherwise. And the same set of challenges in differentiating signal from noise exists when you're trying to detect sarcasm in your deadpan friend's conversation, the tone in an email your boss sent you, the taste of cinnamon in the apple crumble, or a tick in your dog's fur.

WHAT COUNTS AS SIGNAL? WHAT COUNTS AS NOISE?

With these definitions of "signal" and "noise," you may already suspect that one person's signal is another person's noise, since they each may be looking to detect different aspects of the world. For example, imagine that you are watching a movie on TV and during the course of the movie, four things happen:

A. In the movie, a brick flies through the window into the room where the protagonist is standing, causing a loud crash.
B. In the movie, a pea-soup-thick fog makes it difficult for the protagonist to make his way through the desolate woods.
C. The movie is interrupted by an emergency announcement warning you of an approaching wildfire.
D. In the movie, during a dramatic political speech, the protagonist makes an important point about democracy.

Which of these are "noise" to your movie-watching experience? You might be tempted to say B, because a thick fog seems to qualify as a distraction. But in your role as a movie watcher, the thick fog is not noise; it's actually important data that the protagonist is having a hard time getting through the fog. The fog is a signal to you as a movie watcher; it's part of the plot. So we have to choose C. As a movie watcher, the irritating announcement about some sort of wildfire is a needless distraction. Clearly, it's noise to you in your role as movie watcher—although maybe not in your role as a human being who would like to be alive tomorrow.

How many of these are noise from the point of view of the protagonist of the movie? We can ignore C, because the protagonist does not experience this, but now, the fog really is noise. The signal for the protagonist is the path through the woods, and the thick fog is getting in the way of him finding that path. Might A, the brick flying through the window, also be noise? That event is certainly noisy from the point of view of the protagonist, but only in terms of the standard meaning of the word "noise." Thinking of noise as a distraction, the brick through the window doesn't qualify; you've got to assume that for the purposes of the movie, somebody throwing a brick through the window is significant to the protagonist and part of the plot. It's telling him something he needs to know.

How many of these are noise from the point of view of a dog trying to get the movie viewer to take it for a walk? We probably have to go with "all of the above." The dog is not interested in any of this stuff. It is just trying to go for a walk, and anything that happens in the movie or is announced on the TV is just background noise from the dog's point of view, stuff that gets in the way of the important signal, which is the dog's pathetic I-need-a-walk gaze.

Now let's turn the question around and ask whether all four of these events, A through D, could each be signal in some context or other. You might think, No, because event B, the pea-soup-thick fog, would seem to be noise in any context. But suppose you have a flight reservation; the fog is a signal that your plane might be delayed.

The punch line of this particular set of questions is that it's not always immediately obvious what the signal is and what the noise is. You have to consider what you're trying to figure out—what counts as a signal in a particular context, what counts as the noise. It's probably worthwhile, in many situations, to ask yourself several questions: Is there a signal here? Is there a noise here? Are we clearly differentiating them? Could we perhaps be confusing one for the other? This sort of inquiry becomes important when we start looking at real-world problems, such as we find in questions concerning climate change and global warming. Take a look at the graph below, which shows the annual global average surface temperature measurements over the period between 1850 and 2000.[40]

You obviously want to know if there is a signal here of increasing temperature. You may also want to know if the warming signal starts around the same time that we humans started putting carbon dioxide into the atmosphere in much larger quantities. But the first thing you see when you look at the graph is that it's actually pretty noisy, in the sense that the measurements are bouncing up and down a lot from year to year and from decade to decade. You really need to know a lot about what's causing this noise, these variations in temperature that appear to be

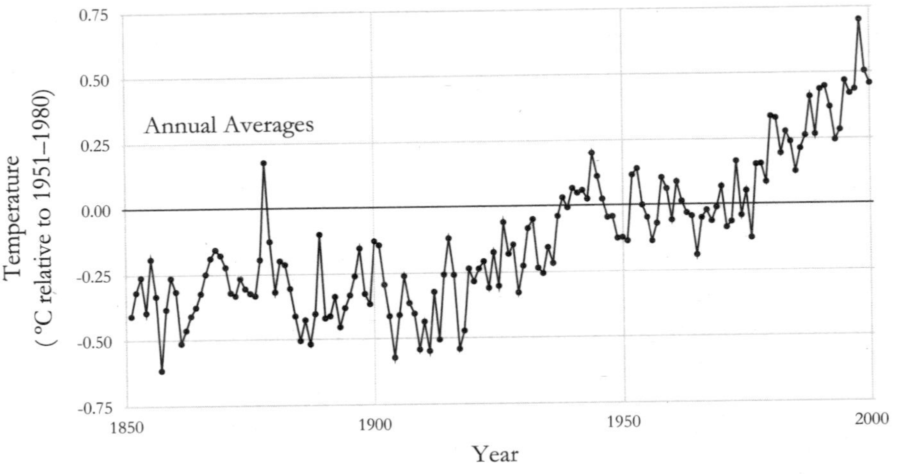

random, in order to be able to interpret the graph and detect the signal. When you have a little rise, say a quarter of a degree or so during a twenty-year period, is that a signal of climate change? Or is it just one of the random fluctuations that you're going to get in this particular system? After all, we see lots of bigger jumps year to year.

The story gets even more dramatic if you look at the fluctuation in global temperature in a longer-term context. In the graph below, which shows twenty-year average temperatures going back to 400 BC (using

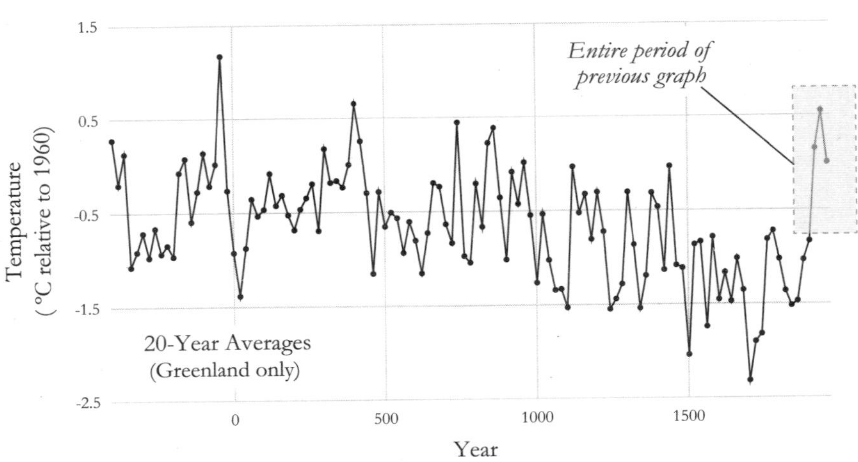

Greenland ice-core data as a rough proxy for global temperature changes), you can see that the period shown in the graph above is the small section over on the right, and it sure looks like it's just part of a long series of ups and downs—that is, noise rather than signal.[41] (Rome was built during a warm period, and the Black Death raged in a cold period; maybe we prefer it warm.)

Unfortunately, as we continue the plot into the more recent years (see the next graph), the signal of global warming appears to rise more strongly relative to the noise. Moreover, more comprehensive study of the recent rise in average global temperature indicates that in the case of the past century, most of the rise is being triggered by things humans are doing. Much of the evidence for this comes from recognizing the sources of the noise in this graph. For example, many of the small short-term drops in temperature are associated with volcanoes that put gases into the atmosphere that cool the earth by reflecting sunlight. We also have indications that the noise isn't just an indicator of poor land-based measurements, but is actually reflecting reality, since most of the remaining short-term fluctuations well match the fluctuations in the surface temperatures of the North Atlantic Ocean. Other than these two short-term noise correlations, the other trend is the slowly rising shape in the plot of temperature that is well matched to the slowly rising shape in a plot of CO_2 gas in

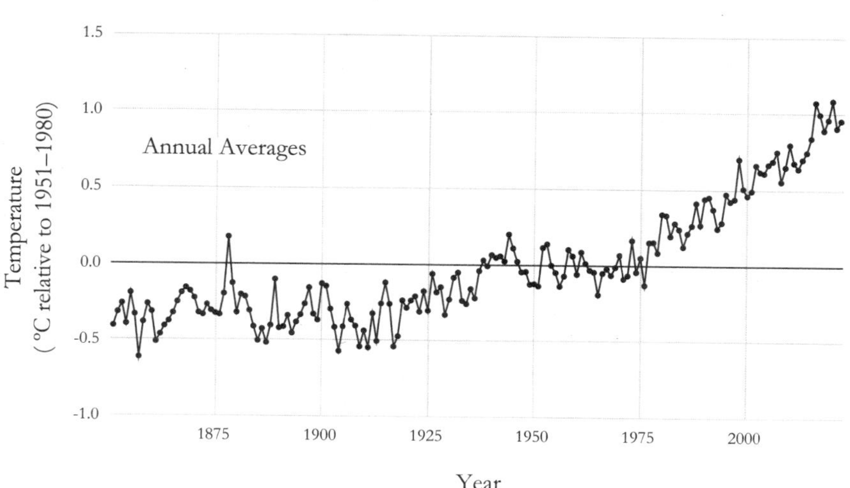

the atmosphere over that same period—which is the signal that we were worried we might find, since the change in CO_2 is primarily human-caused.[42] This is a good example of a problem where separating the signal (the trend of temperature rise associated with rising CO_2) from the noise (the short-term fluctuations associated with volcanoes and ocean temperatures) is key to understanding a situation.[43]

Stepping back to look at the big picture, we see that—just like with the cocktail party effect, where you could, in principle, figure out who else is talking and making it hard to listen to your friend—you can sometimes pin down the sources of what looks like random noise that interferes with the signal. Often, as in a cocktail party, this is not worth doing, since we have other ways to enhance the signal and suppress the noise (for example, we could move to a quieter corner to have our conversation). In the case of global climate, there is an additional reason to identify the sources of noise, however: our desire to keep the earth's temperature in the range of the recent century, if we can, given the consequences of a significant temperature change in a world supporting billions more people than ever before. Controlling temperature would be difficult to do, though, if we didn't know which variables (like CO_2) are the levers for change, and which are simply noise sources that don't affect temperature in the long term.

Once you begin thinking about the problem of detecting a signal amid noise, you start to notice it everywhere as you go about your day-to-day life, not just when you go to a cocktail party. When you read your child Aesop's "The Boy Who Cried Wolf," you realize that the story is a fable about a boy who, by creating noise, loses the ability to send a warning signal. And when you show her the *Where's Waldo?* visual puzzles, you think how cleverly the illustrator creates a pattern of images—the noise—that mimics the rough visual impression of the cartoon character Waldo—the signal.

Watching the news, we become more sympathetic to the people whose job it was in 2001 to detect terrorist threats but who didn't find the signal—of the 9/11 plot—available in the reports of the odd behavior of people training to fly airplanes but uninterested in learning to land—amid the noise of myriad other reports of suspicious behaviors in the world. There is probably a similar story about the missed warning signs of the attack on Pearl Harbor.

And there are even some rare situations in which noise is our friend, when we would *like* to ignore the signal. Most obviously, that's why "white noise machines" have the word "noise" in their name: they help us avoid listening to the signal of the fascinating but annoying conversation going on in the next room when we want to sleep. One of the possible explanations for why we scratch an itch offers a more surprising example of the benefits of noise: it could be that we are masking the annoying itch signal coming from the mosquito bite on our arm with the noise of a less annoying scratching sensation.

JARGON: SIGNAL-TO-NOISE RATIO

There is another bit of scientific jargon conceptually related to the concepts of signal and noise that is probably worth knowing even if it's not something we use day-to-day. The jargon arises because scientists often have to design techniques to pull out a signal buried in noise, and this requires knowing quantitatively how deeply the signal is buried; that is, what exactly is the size of the noise compared to the size of the signal. The term for this is "signal-to-noise ratio."

The basic process of determining a signal-to-noise ratio is easy to demonstrate. Let's say you have a signal consisting of sixteen characters:

A_STITCH_IN_TIME

Now let's imagine that this is a secret message and it is being read to you letter by letter over a poor phone connection with lots of static acting as noise. To mimic what this might look like, let's add noise to this signal, by randomly choosing two of the characters and replacing them with randomly chosen characters:

AQSTITCH_VN_TIME

We can think of the "signal-to-noise ratio" in this example as 14:2 (or 7:1)—fourteen characters of signal and two of noise. At this ratio, or noise level, it is still possible to make out the signal, or at least guess at what it is. What happens if the phone connection is even worse? We simulate this by replacing two more characters with noise:

EQSTITCHNVN_TIME

Now, with the signal-to-noise ratio at 12:4 (or 3:1), it is considerably more difficult to detect the signal. You might guess at it after many tries, but the noise is really getting in the way. If you replace yet two more characters with noise to reduce the signal-to-noise ratio to 10:6 (or 5:3), detecting the signal is nearly impossible:

EQATITCHNVN_TUME

You can probably see from these examples that luck can play a role when a signal gets hidden by noise. Some kinds of random information do a better job of hiding a signal than others. In science, you sometimes end up just being lucky. Some signal comes in, in a way that's fortunately not too badly distorted by noise, and it gives you a hint that leads to a discovery.

The main point, though, is that the quantification of the signal-to-noise ratio helps us measure noisiness and signal quality and compare situations. In our example, a 7:1 ratio leaves the signal fairly evident, but at a ratio of 3:1, things get pretty dicey. With quantification, you can actually project what level of signal versus noise you're going to need for your purposes. This is something that scientists have to do (or should do) often, so it's worth knowing that this is frequently an important consideration when a scientist is asked to opine on a key factor in a decision.

Now, when scientists talk about signal-to-noise ratio, they often use a fancier statistical definition of what that ratio is. But the same principle applies: you're trying to figure how much noise you have compared to how much signal you have, and what the odds are that, given this ratio, you'll be able to recognize the signal in that noise. This is why the term "signal-to-noise ratio" is important, useful jargon to know.

WHAT TO DO ABOUT NOISE

We mentioned that scientists are often in the business of pulling out a signal buried in noise. Let's look at one of the ways this is done, using the example of a fictional scene from a dramatic movie with you as the star (on the theory that it's more fun to star in our own fictional movies). Imagine now that you're on patrol, flying your World War II plane out over the Pacific Ocean. You find yourself picking up a radio signal that sounds like static. What do you make of this roar of noise that's coming in over your earphones?

Since you've been trained in such matters, you have a theory. You think maybe there's a signal that's coming in at a certain pitch, at a certain frequency, and it seems you hear it ever so faintly. It turns out that your radio has an equalizer on it, a frequency filter, so you can suppress all the other pitches that are not the one you're trying to hear. You fiddle around with the radio a bit and then you hear it: three short pulses, three long pulses, three short pulses. It is SOS in Morse code. You turn your plane in the direction of the signal to investigate.

Now, there are two interesting aspects to this story. First, the only difference between hearing that wash of noise and hearing the signal was suppressing all the frequencies that were not where the signal was. You filtered out everything else, and the only thing that came through was that one frequency carrying the signal. Suddenly, it became obvious there was a signal there.

The second interesting aspect is that it turns out that our brains do that job of filtering out noise really, really well—if they know where to look for the signal, in this case the pitch at which the SOS signal is expected to be broadcast. In fact, if you were that World War II pilot who had just heard an SOS signal hidden in lots of static and you turned that equalizing filter off, you'd still be able to recognize the signal despite all the noise, because you'd know where to listen for it, and your brain would act as the filter. The brain is amazingly good at learning to filter. (That's important to keep in mind as you are looking around for signals in noise, since we are about to explore the problems that arise precisely because our brains do this filtering automatically. As with all of our human gifts, there is a flip side.)

A classic example of looking for signal by filtering the noise is the ongoing international scientific effort in the search for extraterrestrial intelligence (SETI). One of the things you do as a SETI scientist is point a radio antenna out to distant stars. What you hear is static that sounds a lot like what the World War II pilot was hearing. It's the static roar of all the noise in the universe—or rather the noise from that particular location up in the sky where you've aimed your antenna.

The problem is that you can't just listen for beep-beep-beep-beep sounds, because as far as we know, the ET doesn't speak in Morse code. So what you're trying to do is invent all the possible filters that you can imagine. That's what the SETI scientists' job really is: not to just blindly look for an obvious signal in the noise, but to invent filters that might focus our attention on the communication signal that they think an extraterrestrial intelligence might use. It's not obvious how to do this. If you think about it, what communication signal would you want to focus on (and filter out everything else)? What do you think an ET would use for its equivalent of Morse code?

One simple possibility would be to see if they're trying to send a steady repeated pulse at a certain pitch. In fact, it would be interesting if it turned out there was a specific pitch to look for, as we found in the SOS example. Unfortunately, there are other natural phenomena out there in the universe that produce repeated pulses that are not extraterrestrial intelligences.

We will come back to a dramatic example of this when we finish our opening cliff-hanger concerning the planet orbiting a pulsar, but first we should discuss a particularly insidious problem we face when we are playing this game of filtering for a significant pattern in the midst of noise:

our brains see patterns in random noise—and attribute meaning to the patterns! As we'll see, our ability to recognize the signals we need for our everyday (and long-term) decision-making amid all the sources of noise depends on how well we understand our propensity to be fooled in this way. This is such a major topic for how we all should work with signal and noise every day that we now turn to the next chapter to give it—and that pulsar story—their due.

CHAPTER 7

Seeing Things That Aren't There

The next part of the story starts with our naive expectation of what random noise can do.

In what sense are our expectations naive? We did this experiment: First, we asked students in a class to flip pennies fifty times and record heads with an "H" and tails with a "T." Then we tasked another group of students with creating a sequence of "heads" and "tails" that would seem to them as random as flipping coins, without actually flipping any. Here are the sequences they came up with—though we won't tell you yet which sequence, A or B, was generated by actually flipping coins:

A:
HTHTTHTTTHTHHHHTTTTTHHTHHTTTHTTTTTHTTTTTTTHTHHHTHT

B:
TTTHTTHHTTTTTHTHHTTHTTTHTTTTHTHTHHHTTTHHHHHTTHHHTH

Can you tell which sequence is truly random, and which is just intended to appear random, just by looking at it? When you compare the two sequences, your brain immediately pays attention to the runs of "H"s and the runs of "T"s (emphasized by surrounding rectangles). Sequence A has a run of seven "T"s, and two runs of five "T"s, an amount of "runniness" that's not matched by sequence B—although sequence B does include a run of five "H"s and a run of five "T"s. On this basis, you might be tempted to say that sequence A is the fake one, because a "true" random sequence wouldn't be likely to have so many runs, including seven "T"s in a row.

If so, you'd be wrong. Sequence B is the fake. The students who created it probably thought that it wouldn't look random if they put down too many tails in a row or too many heads. (They did try to put two longish runs in, but then lost heart.) After going "Heads, heads, heads, heads," or "Tails, tails, tails, tails," they probably thought, This doesn't feel random, and cut the runs short. But real random sequences have lots of surprisingly long runs that we don't expect.

Now let's put this in the context of our previous examples of a WWII pilot trying to find an SOS signal in a wash of static, and of the hunt for a sensible phrase after putting random wrong letters in the midst of a string of words. In both cases, we were aware of how hard it can be to find signal in noise. The new element of the story that you should be aware of is that when you're hunting, hunting, and hunting for a signal in what looks like very noisy data, you will get fooled. The noisy data will eventually show you things that look like meaningful patterns, and it will do that in ways you never would have expected. In other words, if you hunt for a signal in random noise, you often think you see a signal when in fact there isn't one.

None of us really has a good feel for how often patterns appear in random noise. Perhaps the best we can do is learn to always question our pattern finding and know that our intuition about what is a signal is likely to be wrong—and requires comparison to how often that pattern appears in random data. (Statistics offers us many mathematical techniques for doing such a comparison.) If you come away from reading this book with that thought deeply internalized, it will stand you in good stead. So if you have never done so, go find a coin (they do still exist), flip it a hundred times, and write down all the long runs that you find. (And those of you who already know this, go find someone else to amaze and amuse with this demonstration.)

If you are a business owner, you will never again overorder stock based on what looks like a pattern of sellouts five Thursdays in a row, without first consulting a statistician to tell you what the odds are that this is just a random coincidental series of Thursday sellouts.

FINDING THE HIGGS PARTICLE

Okay, what are ways to see this in action—this differentiation of signal from noise? Let's begin with a dramatic, larger-than-life example, the particle physicists' discovery of the Higgs Boson (a subatomic particle), because it shows the extraordinary lengths that scientists have to take to avoid confusing noise with signal. (The next time you hear a reporter trying to get a scientist to be less cautious in their statements, you will have more sympathy for the scientist's carefulness!)

This Higgs discovery was the result of a classic particle physics project, where some well-understood elementary particles—in this case protons, the nuclei of hydrogen atoms—are accelerated to immense speeds by sending them around and around a huge ring, miles across, steered using numerous electromagnets. The protons in each beam are sent around as "bunches." Then one bunch of protons traveling clockwise around the ring is aimed at another bunch of protons traveling counterclockwise, with the goal of creating collisions of individual protons from each bunch that have sufficient total energy to create new elementary particles never before seen. The result of each of these collisions is messy, as the two colliding protons are completely transformed into a rich collection of different particles spraying out in all directions, each taking up some of the total energy from that collision. Almost all of the particles in this spray are well-understood, previously studied particles like electrons, photons, and muons, but by studying the results of trillions of these collisions, particle physicists can look for the rare cases in which a never-before-seen particle is found. For the Higgs experiments, scientists were looking for a particle that had been predicted to exist, called the Higgs particle, which could be recognized by having a unique mass, different from all other known particles.

You have to imagine you're at the particle accelerator in Geneva, the

Large Hadron Collider—the LHC. You go down in an elevator, 100 meters deep into a tunnel, and you find yourself in a huge underground ring 27 kilometers in circumference, lined with thousands of electromagnets, each of them longer than a bus. At several points around this underground ring there are cavernous chambers at the locations where the bunches of protons are collided with one another, and gigantic detector systems have been built at each of these locations by different international collaborations of scientists. The detectors are comprised of many smaller parts ("subdetectors") designed to capture the signals produced by each of the particles that spray out from the collisions.

The goal is to use all of this data about the signals to figure out what new particles the collisions have produced. In particular, it's important to calculate the energy contained in these particles, since this can tell us something about the energy that was contained in hoped-for new particles, like Higgs particles, that were temporarily created in the collisions. (Here's where everybody who has grown up in the twentieth or twenty-first century recites our favorite equation from Einstein, $E = mc^2$, which tells us that if we know the speed of light, *c,* we can convert that total energy, *E,* into the total mass, *m,* of the original particle—but you knew that!) The scientists collect statistics over many, many thousands, and millions, or even trillions of these collision events, and they build up histograms (data graphs) showing how many of these collision events yielded evidence (from the resulting spray of particles) of the presence of a new particle with a given total mass.

Two major international collaborations, each with its own detector system in one of the collision chambers, were competing in the search for the Higgs particle. The graph below shows the data from the Higgs search at one of those two detectors, the ATLAS detector.[44] (We authors were all teaching at Berkeley while the inner part of this detector was being built there—so the ATLAS team was something of the "home team.") What you see in the graph is what looks like a little excess, a little bump, that represented this exciting discovery, potentially, of a Higgs particle where they were expecting that they might see it, because it would have a mass somewhere around the mass indicated by the downward-pointing arrow.

The Higgs particle is a big deal, because if we find it, we finally have convincing evidence for a theory that explains how most of the stuff around us gets to have mass (unlike all those massless photons that go whizzing around us all the time)—and we have been waiting for this evidence for well over forty years, since Professor Higgs (and several other scientists) first proposed this explanation.

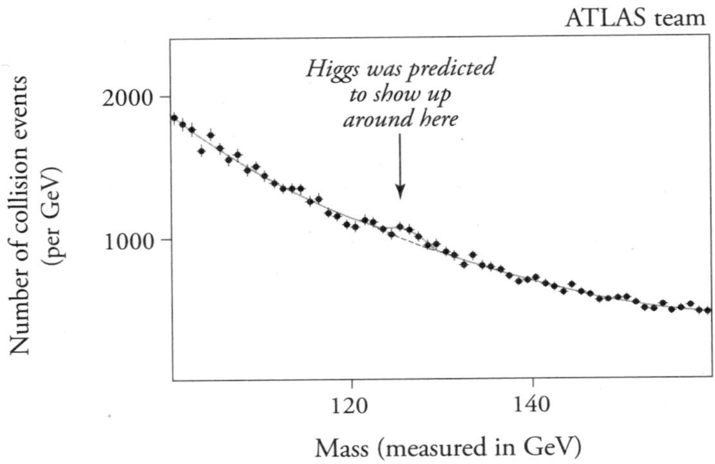

Now, the billion-dollar question—quite literally, since they spent something like $10 billion to build this experiment—was, is this little bump in the plot real? How do we know that this is actually a signal and not just a noise fluctuation? If you look at the graph, you can see other places that are a little high, too. If this particular bump were not emphasized by the solid line drawn through it, and the arrow pointing at it, would you be sure that it was different from the other points on the graph that just happened—due to random noise—to be a little bit higher than the others?

By now, particle physicists have been in this particle-finding business a long time, so they saw this problem coming when they were building the LHC. They quite sensibly had *two* teams build detectors at two of those different collision chambers in the LHC accelerator ring, which meant that their two competing results could be compared with each other. Since so much depended on that comparison, the teams agreed that they would come out with their results at the same time, and not try to scoop each other with inconclusive results.

The CMS (Compact Muon Solenoid) Collaboration (or "the other team," as we at Berkeley called them) therefore had their own version of that same histogram plot, and everybody was excited to compare the two teams' results.[45] This is what the CMS team saw:

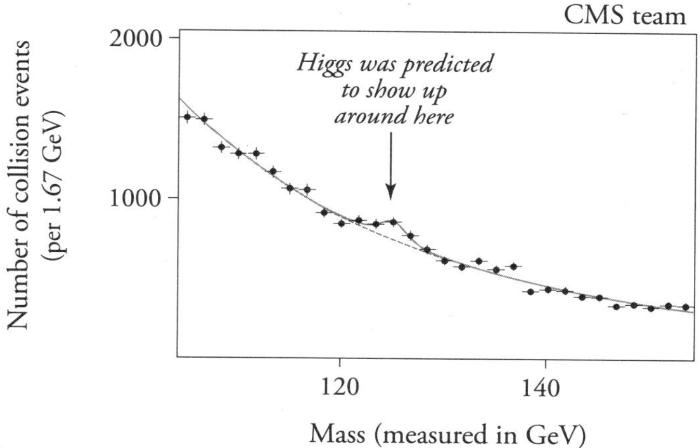

So, at the other side of the ring from ATLAS, the CMS team was seeing a similar little bump in their data. With both teams seeing this bump, the scientists could make a claim that they were seeing what they called a Higgs-like particle. (They were still being cautious and calling it Higgs-like rather than the Higgs.)

This is the name of the game for particle physicists—and very often for us in real life: seeing a signal in the noise and trying to be sure it's not a fake signal. It's so important that trying to get good at not being fooled by a little fluctuation in the data is part of the culture of science. Physicists will play all sorts of games with one another to train in this skill. They'll show data to a colleague and ask if they think a bump on the graph is real or just a random fluctuation. For example, when they see a new result like what they had with the Higgs, they often try to generate lots of examples of what similar data would look like coming from their detector if there were no Higgs particle (so all the bumps on the example graphs are only due to random fluctuations), and then they'll hand you ten example graphs with the real graph mixed in and ask you to identify which ones are the random data and which one is the real signal. And if you end up choosing the wrong plot, then it's obvious that what you thought was an interesting signal is just as likely to be random noise.

This extreme care to distinguish real signals from noise makes sense for physicists. They are striving to identify fundamental aspects of the world, rules, and entities that are always going to be in place no matter what we do, so that we can build all of our predictions and technologies on these deep truths about reality. This explains why the reporters we mentioned earlier have such a hard time getting the scientists to overcommit about their findings: The scientists want to be sure that the new element of reality they think they have identified will stand the test of time. So they would only say it was a "Higgs-like particle," not the discovery of the Higgs!

For most of our day-to-day hunts for signal in noise, we don't have to be quite this careful, but once you get sensitized to the issue you start to realize that we, too, are playing this game of finding signal in noise all the time: Did the bus driver just announce the stop you are supposed to get off at as "Eastham Racetrack," or did he say, "Please move to the back"? Did the bicycle rider up ahead of you just wobble around a bump in the road, or are they moving into your lane? Is that a stain on the sleeve of your shirt or just the way the light is falling on the bunched-up material where you rolled up the cuff?

Similarly, we are always trying to read one another's intentions through all the noisy clues of one another's behavior. Is my date getting serious? Who in this room liked my suggestion? More tragically, our governments have not been able to recognize real threats among all the reports of worrisome behavior—for example, the missed warnings before Pearl Harbor and 9/11 that we mentioned. In situations where a lot is at stake, like these intelligence examples or in business investments, we start to need almost as much caution as the physicists do.

MORE NOISE, MORE CHANCE OF BEING FOOLED

Of course, most of the time, we don't get to compare results with a second detector like the Higgs search did (even though most signals we are interested in don't cost billions to detect). So we have to get good at recognizing the situations where we are particularly prone to imagining we see signal when all we have is a noise fluctuation. With that in mind, it's time to add the next level of complication to the story: the more data you look at, even data that consists of just noise, the more examples of surprising patterns you'll find—so you will think you have found a signal. It is one thing for a single individual to flip a coin a dozen times, and quite another for a group of people to be flipping coins all day; in the latter case, you can get some very interesting results.

Let's say a friend tells you that when conducting coin flips, she sees ten heads in a row. This, of course, is a surprising result. But in which of the following contexts would that result be the most surprising?

A. She is the only person flipping a coin, and she flips it only 10 times.
B. She is one of many of your friends flipping coins, and everyone flips 10 times and reports back to you what they find.
C. She is one of many of your friends flipping coins, and everyone flips 100 times and reports back to you what they find.
D. Your friend is the only person flipping a coin, and she conducts 100 flips.

The answer is A. If you think about it, the odds that one of your many friends will see ten-in-a-row heads if a lot of them are flipping coins are much higher than if just one of them is flipping. We can argue particularly strenuously against C, because if there are many people flipping and each one does *100* flips, getting 10 heads in row isn't all that surprising. (In fact, if you have fifteen friends, the odds are more likely than not.) This may seem to be stating the obvious, but it's good to be sure about this starting point.

Now, the reason that this gets hard is that you don't always know how many "coin flips" were done when somebody publishes a new scientific result, or when you read about some surprising bit of news. In fact, often even the people who actually did the research don't realize how many "flips" they've done—that is, how much opportunity their data gives them for seeing patterns that appear nonrandom. A couple of examples may make this clear.

First, the stock market. When you want to invest in the stock market, as a typical small investor, you quickly realize that there are many companies out there that have a stock portfolio expert who puts together a mutual fund, and they want to sell you shares in their fund. To do that, they will say that their mutual fund has outperformed the market in the last five years and that they've predicted the market correctly multiple times. You might then buy that particular mutual fund, thinking that its manager is better than other mutual fund managers. The graph below shows the performance of the top mutual funds for a five-year period from 2013 to 2017. It is ranked according to how good each mutual fund manager was at calling the stocks. You can see that some did 4 percent better than average, many did about average, and then some did about 10 percent worse than average. Looking at this, you might be tempted to

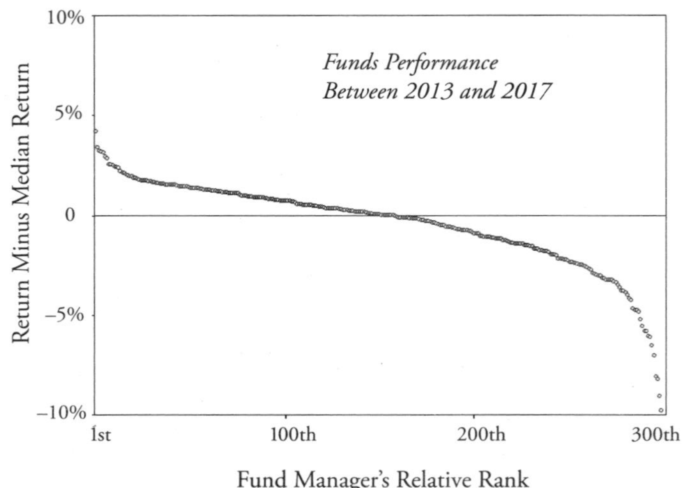

believe that the portfolio managers with the best performance records were in fact more skilled or more knowledgeable than those with the worst records.

What happens when you measure the performance of those same people for the next five years, between 2018 and 2000? What you find is that it's almost completely random, as you can see from the graph on this page. Performance in one five-year period has virtually no correlation with performance in the next five-year period. For one five-year period, a few people were really good with economic analysis and with gut feelings. However, for some reason, those same gut feelings and analysis tended not to work over the next five years. What we're seeing is that if you study enough fund managers, eventually you'll find somebody who seems to have called all the right stocks—but this will just be the effect of seeing patterns in random noise, with lots of noise to search through. And the interesting thing is that those people are not intentionally faking it. They're really pretty sure that they did very careful study and actually figured out the performance of these different companies and are not just flipping coins. They think of themselves as serious scholars of economics and the fundamentals of good business practices. And yet, apparently, the pattern is essentially random.

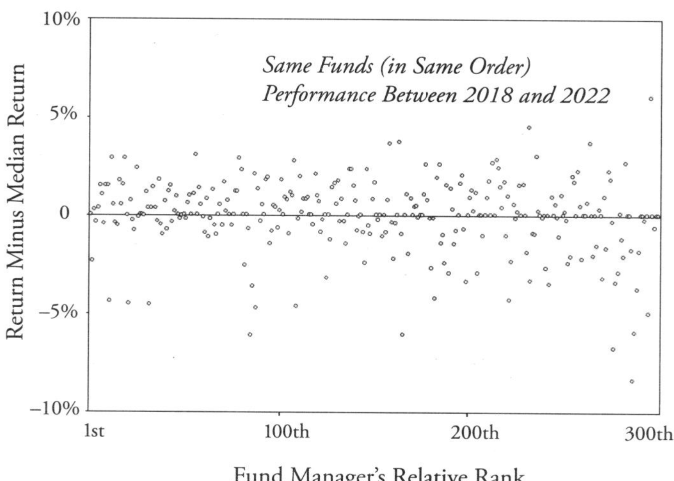

There may be small exceptions—fund managers who almost always do better than average over time (or those who always do worse)—but you should take this as a highly cautionary note.[46] Most people are probably not aware of how many funds are playing this game at any given moment, such that *some* fund will—by luck, not by skill—be able to make all sorts of great claims at any given moment. In any random distribution of fund performances, some have to be doing better and some have to be doing worse. (This is the underlying reason that it is often advised to stick to a big index fund, and not waste money paying a fund manager's fees.)

THE LOOK ELSEWHERE EFFECT

There is another route to mistakenly thinking you see signal when you have simply been fooled by a pattern in random noise. This mistake, called the look elsewhere effect, occurs when instead of hunting for one pattern in a noisy dataset that could indicate a signal, you hunt for several different possible patterns, upping the odds that random noise will show *some* hunted pattern. Here's an example: Let's say you set up a medical study where you're going to test the hypothesis that taking an aspirin every day is good at preventing a heart attack or at least diminishing the risk of heart attack. You enroll a thousand people, and you begin to run that study, but since you've gone to the trouble of enrolling people and doing that study, you find yourself thinking that it might be good to also test if the aspirin taken every day helps reduce the risk of cancer. And then you think that while you're at it you should check the effect on lung cancer and colon cancer specifically. And maybe asthma, too. Now you're beginning to "look elsewhere." You're looking at many more variables than you had originally designed the study to test for.

Now imagine this taken to the extreme—you are looking at the effect of a daily aspirin on a hundred different diseases and you still have the thousand people in your study that you started with. You can start to appreciate that the odds of getting something that looks like seven-in-a-row heads goes up, because now there's more opportunity for rare chance events to falsely make it look like aspirin has an effect on at least one of these many diseases. This is equivalent to doing many, many more coin flips. You will start seeing things that look like real effects. And this does happen in medical research. In fact, rather visible examples of this have occurred in studies of hormone replacement therapy, and studies of effects of lead in drinking water, for which results were later questioned because the researchers had looked at too many variables without setting up the studies and planning their analyses for that number of variables.[47]

Now, of course there are ways to compensate for the look elsewhere effect. Before you ever look at your data, you can precommit to studying the variables that you're going to be testing for and calculate the number of subjects you need for that number of variables. So it's possible to design a study where you can look through more variables, but that requires more subjects in order to have the statistical power to get valid results.[48]

All this may help you better understand the challenges the physicists faced when examining their data for evidence of the Higgs particle. When looking for the Higgs, physicists did not know exactly what the mass was—they did not, in fact, have that nice arrow pointing to the correct place on the graphs that we have shown. So they would have been willing to consider any suspiciously high blip over the entire range of energies shown on that plot as being the sign of a Higgs; they were certainly looking elsewhere than just at the location of the arrow on the plot. Of course, particle physicists have long-ago recognized this problem, so they set a much higher bar for what would count as a real peak, even before they took the confirming step of comparing the results from the two detector teams. (Back to the case of trying to determine if your date is interested in you: If you are going to consider every possible bit of body language as a possible source of evidence, then you'd better hold out for pretty strong versions of those indicators, since otherwise you would be overinterpreting every random move they make: "Oh, look, one of their feet is vaguely beginning to point in my direction—and is that their left pinky raised to indicate happiness?" You should really hold out for long, intense eye contact, not just a passing glance.)

PULSAR EPILOGUE

When we last left our heroes (the team Saul was working with as a postdoctoral student) in Chile, exploring the remnants of Supernova 1987A, they were seeing what looked like a new pulsar signal with wandering pitch, and when they made slight corrections for the little Doppler shift that came from our Earth spinning and going around the sun, the "wandering" when plotted on a graph turned into a beautiful sine wave that looked just like what they would see if a planet was orbiting their supernova remnant. At that point they wrote up an article on their work and sent it to *Nature.*

The following year, the team went back to observe what had happened to the pulsar in the time since they'd last been there. They were expecting that the frequency of the pulse would be slowing down; that is, that the "pitch" would be dropping, because pulsars give up a lot of energy in spinning a couple thousand times per second in the form of gravitational waves. They got back to the telescope, but the signal didn't appear. The next night, they went back again, and there it was—the pulsar signal was back. And then not so much the next night. At some point the team started realizing that there was a correlation between seeing the signal of the pulsar and an instrument across the other side of the telescope dome that was used for another purpose: whenever that instrument was on, the pulsar was there; when it was off, no pulsar!

The instrument that the team was using to detect the pulsar[49] was a highly sensitive photodetector system—so sensitive that it ended up picking up a signal leaking from this other piece of equipment that happened to be broadcasting at just about the frequency they'd expect from the pulsar—about 2,000 times per second. And the signal from that other piece of equipment happened to drift, coincidentally, in such a way that if you corrected for the spin of the earth and our motion around the sun, it turned into a beautiful, perfect sine wave. This was noise doing something that looked like signal. It meant, of course, that the next month, instead of writing papers, the team was writing a retraction to the article in *Nature.* They had not, in fact, discovered the first evidence of a planet outside of our solar system.

This is the kind of story that should strike fear into the heart of every reader. Whether or not you plan to be a scientist, you are likely at some point in your life to be putting serious money down on the odds that something won't miraculously go from messy noise to a beautiful signal just by random chance. And every now and then that bet will turn out to be wrong, because the statistics of random numbers will eventually give you a surprising thing that looks like a signal.

Saul comments, "Fortunately, in this case, I was young amid a crowd of very experienced and distinguished other scientists, and I think that we were fast enough to explain what was going on that people did not hold it against us too much. But I will say that this is the kind of error due to misinterpreted noise that was very clearly in my mind when a dozen years later my team found evidence that the universe's expansion is accelerating. You can see that if you've been burned once by random fate, you'll be very cautious thereafter. We really had to think very hard before we went public with the result about the accelerating expansion of the universe."

In the end, this later team working on the universe's expansion tried every cross-check and test they could invent before they announced—with a quantified confidence level!—their astonishing measurement indicating that the expansion is getting faster and faster, perhaps due to a previously unknown "dark energy" that dominates the behavior of the universe. Unlike with the putative pulsar, this result has since been backed by multiple other measurements, some using the same measurement technique and others triangulating with completely different approaches. Today, the hunt for an explanation of the accelerating universe and a possible dark energy is one of the key questions of modern physics.

In all of the above cases, scientists have to judge whether they have gathered the right quantity of data to give them sufficient confidence to conclude that they have found a signal rather than noise. And in the examples from our daily lives, we similarly need to be able to gauge the odds that what we think is a real pattern isn't just random. How do we decide what odds are good enough, so that we can tentatively claim that we have discovered a signal? That is the subject of the next chapter.

CHAPTER 8

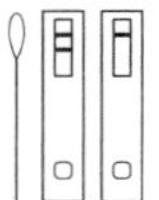

Pick Your Poison: Two Kinds of Error

There is a tension between the data we collect and the real-world decisions we need to make. Imagine again we are jurors in a criminal trial. We need to sift through the trial evidence to judge whether the defendant is guilty of the crime. The defendant might indeed be guilty (i.e., there is actually a true signal present) or the defendant might be innocent (i.e., the prosecution's evidence is noise, not signal). In a criminal trial, jurors receive a lot of evidence, some of it seemingly incriminating and some of it seemingly exonerating. Thus, any signal is mixed in with a lot of noise, so our judgment of guilt will be probabilistic. But if we need to make a decision, we can't wait for 100 percent certainty; we need to decide, "Do I have enough evidence to justify my choice?" We do this by adopting a *standard of proof*—a threshold level of evidence we will require before we reach a conclusion. At both criminal and civil trials, the judge will instruct us how to do this: We should only convict the defendant if we believe the prosecutors have met the evidentiary standards and burden-of-proof requirements required by law in those proceedings.[50] In other situations, no one instructs us on what standard of proof to use, but anytime we make a categorical (yes-no) decision based on probabilistic evidence, we are applying a standard of proof, whether we realize it or not.

Setting a standard-of-proof threshold requires us to balance two kinds of errors. For example, in the criminal trial case, the defendant either did or did not commit the crime (true state of the world) and we can find the defendant "guilty (convict)" or "not guilty (acquit)," creating four possible trial outcomes:

		Trial Verdict	
		"Not Guilty"	**"Guilty"**
True State of the World	**Committed the crime**	*Set a guilty person free*	*Convict a guilty person*
	Did not commit the crime	*Set an innocent person free*	*Convict an innocent person*

As conscientious jurors, our verdict choice is thus fraught with risk; there are two good outcomes (where "the system works") but also two kinds of grave errors. It is horrible to contemplate punishing an innocent person, but also, if the crime is not minor, distressing to consider that we might set a dangerous person free to commit other crimes.

The criminal trial is a familiar setting for most readers (from movies and TV if not from personal experience). But we struggle with similar dilemmas in many domains of life. For example:

Should I show up at the airport early to guarantee I don't miss my flight, or should I arrive right before departure to avoid wasted time in the airport lounge?

Is it better to be permissive about my teenager's social life, enabling her to feel trust and to develop autonomy? Or is it better to be restrictive, minimizing the chance she'll end up in harm's way?

Should we enforce rent control to help poorer people stay in their apartments when housing prices rise, or is it better to let prices rise to create an incentive for developers to build new housing units?

Should we let war refugees settle in our country, to provide for their safety and well-being? Or should we ban them because there's a risk that some of them are terrorists or criminals?

One of the ways that science can advance our problem-solving ability is by abstracting away (temporarily) the unique details of a problem so we can focus on general features common to many problems. Thus, we can generalize the criminal trial table to a more generic template:

		Decision	
		"Signal Is Absent"	**"Signal Is Present"**
True State of the World	**Signal Is Present**	*False Negative*	*True Positive*
	Signal Is Absent	*True Negative*	*False Positive*

In this revised table, we are going to make a decision based on our best judgment as to whether some signal is either absent or present in the world. This "signal" can be any dichotomous state: innocence versus guilt in a trial, undeserving versus deserving of government assistance, ordinary storm versus tornado, no cancer versus cancer, and so on. If we decide the signal is present, we call our decision a positive; and if we decide it is absent, we call it a negative. This creates four generic types of outcomes: If no signal is present, we can say "absent" and get it right (a "true negative"), or say "present" and get it wrong (a "false positive"). If the signal is present, we can say "absent" and get it wrong (a "false negative") or say "present" and get it right (a "true positive"). The tension between false negatives and false positives can also be thought of as "sins of omission" versus "sins of commission."

One immediate implication of the table is that it is not meaningful to simply label our decisions either "accurate" or "inaccurate," because there are two different ways of being accurate and two different ways of being inaccurate. In fields like medicine (where the signal might be "cancer") and educational testing (where the signal might be a correct answer on a test), there has been a shift away from reporting overall accuracy rates (i.e., the proportion of all decisions that are correct) toward more informative criteria about that accuracy—"sensitivity" (the proportion of times we say "present" when the signal really is present) and "specificity" (the proportion of times we say "absent" when the signal really is absent).

Any test intended to help us make predictions—whether it's a screening test for cancer or the doctor's board exam—should have high sensitivity but also high specificity. Notice that we can maximize sensitivity by *always* saying "The signal is present"; e.g., if we always say the tumor is malignant, we'll never overlook any cancer. But in doing so we are likely to lower our specificity rate—we're basically calling *any* tumor "cancerous," and the diagnosis becomes meaningless. A useful test threshold needs to strike a balance between these diagnostic criteria.

STANDARDS OF PROOF AND ERROR TRADE-OFFS

Again, in a world full of noise and uncertainty, we are bound to make errors. Often, we care more about one type of error than another. If we'd feel greater regret about a false-negative error (we would hate to overlook any cancerous tumors), we can set our standard (or decision threshold) fairly low—a bias toward saying "Signal present." If we'd feel greater regret about a false-positive error (we'd hate to scare someone who doesn't really have cancer), we can set our standard (or decision threshold) fairly high—a bias toward saying "Signal absent."[51]

In the common law tradition, we instruct jurors to use a standard of proof that is biased against convicting innocent people; the British jurist Sir William Blackstone famously argued that it is better to let ten guilty people go free than to convict one innocent person.[52] Thus, jurors are typically instructed that they should vote to acquit the defendant unless they are certain "beyond a reasonable doubt" that the defendant is guilty.

To some, this might seem to reflect a "soft-on-crime" position, but there are sound reasons for this bias. First, criminal cases pit an individual citizen against the full power of the public prosecutor's office, which typically has far greater resources than the defendant. But second, in many cases (e.g., "whodunit" crimes where we know the offense occurred but not who committed it), there is a logical asymmetry favoring a stringent bias against conviction: if we convict an innocent person, it is likely that we are *also* letting the true offender go free.

Unfortunately, the phrase "beyond a reasonable doubt" is notoriously vague. The judge doesn't tell you whether your doubts are reasonable or not. Jurors have to figure that out for themselves. In one survey of federal judges, about a third of them said reasonable doubt means that you are 95 percent sure, about a third said 99 percent, and a third gave other numbers. One of us (Rob) has attempted to estimate the thresholds that jurors actually apply, and they appear to fall well short of 95 percent.

Unfortunately, the wiggle room created by this vagueness leaves the door open to juror prejudice. One way we've demonstrated this is by giving our students a hypothetical legal case to judge. Half are given a written description of a criminal case that includes this critical piece of information: "The defendant, a twenty-one-year-old Cal student, has been accused of committing an assault in the parking lot of a local bar." The other half are told that the defendant is "an unemployed twenty-one-year-old Berkeley resident." We ask the groups about the probability of the defendant having committed the crime. We find that the students who believe the defendant is an unemployed Berkeley resident are more likely to think, with the exact same evidence, that the defendant is guilty of committing the crime. In essence, they are more willing to give a fellow student "the benefit of the doubt."

Consistent with this interpretation, when we ask students to rate how bad it would be to convict an innocent person, or to acquit a guilty person, the students who think the defendant is an unemployed Berkeley resident are more concerned about falsely convicting a fellow student than about falsely convicting an unemployed resident. This kind of bias has been found in other studies, where jurors' standards are influenced by the defendant's race, physical attractiveness, and other traits—even when there is no logical connection between these attributes and the facts of the crime.

EXAMPLE: STANDARDIZED TESTS AND COLLEGE ADMISSIONS

To further illustrate the interplay between the choices we make and the errors at risk, let's consider the relationship between college admissions criteria and student performance. For many years, American colleges and universities required student applicants to take a standardized test of their knowledge and cognitive ability, e.g., the Scholastic Aptitude Test (SAT) or the American College Testing (ACT) test. In the following figures, we examine the relationship between standardized test scores (our predictor) and student academic performance in the first year of college, each scored from 0 to 100.

In the chart below, each dot represents an applicant to a college. (In actuality, there'd be hundreds or even thousands of applicants, so you can imagine each dot representing 100 students who have identical scores.)

Our chart is segmented by a horizontal line, which is the college's criterion for success; students above the horizontal line are performing adequately and students below it are on "academic probation" and at risk of being expelled. There is also a vertical line, which represents the cutoff score the admissions committee ordinarily uses to decide whom to admit. *But in this one year, the school will admit everyone, so they can see what would happen to the students they would've rejected.* These lines create our

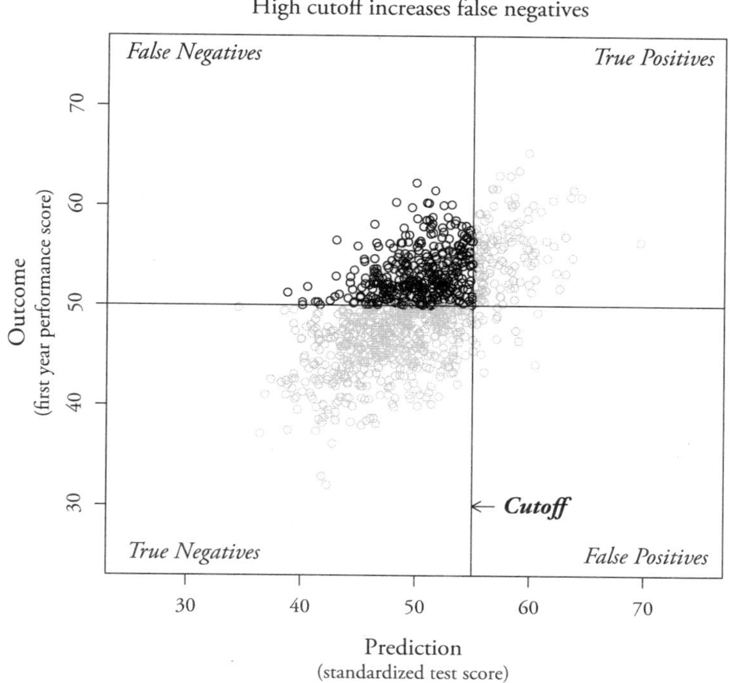

now-familiar 2 x 2 table of two kinds of successful prediction (true positives and true negatives) and two kinds of failed prediction (false positives and false negatives). Notice that the dots roughly form a diagonal line from bottom left to top right. This indicates that there is a positive association between our predictor (SAT score) and outcome (success in college), although it is far from perfect.

Notice that by using a high cutoff score, we will make relatively few false-positive scores—we'll minimize the risk of admitting students who will fail at our college. But we are making lots of false-negative errors—failing to admit students who would have thrived at our college.

Let's say a new college president announces that she wants to give more applicants a fair shot at access to higher education. In the next chart, we show what happens when we lower the cutoff score. Now we make fewer false-negative errors, but at a cost—we're now making a lot more false-positive errors, admitting students who are likely to struggle to get acceptable grades.

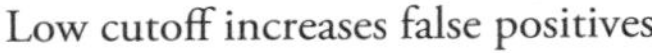

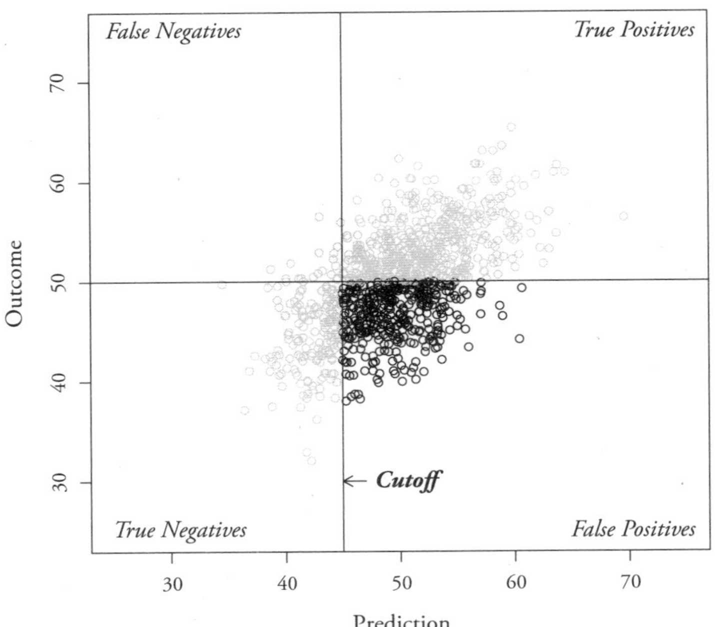

Unless we have a well-calibrated crystal ball, the moment we set cutoff scores on the predictor and on the outcome, we are inevitably going to make errors in admitting only the students, and all the students, who will thrive at our school. Setting those cutoff scores is not a scientific or mathematical calculation; it is a *policy decision* that reflects our values about which error types we'll find acceptable.[53] Real-world decision-making can be informed by data and mathematical principles like signal-detection theory, but they inevitably involve value judgments. And there's no particular reason why scientists are better equipped to make those trade-offs. Such value judgments require some combination of the value inputs of stakeholders—university administrators, faculty, potential students, and their families.

If we set a high cutoff, we'll make fewer false-positive errors but more false-negative errors. If we set a low cutoff, we'll get the opposite result. Does this mean science can't help us with this dilemma? Not trivially. In the next figure, we simulate what can happen if, instead of just adjusting

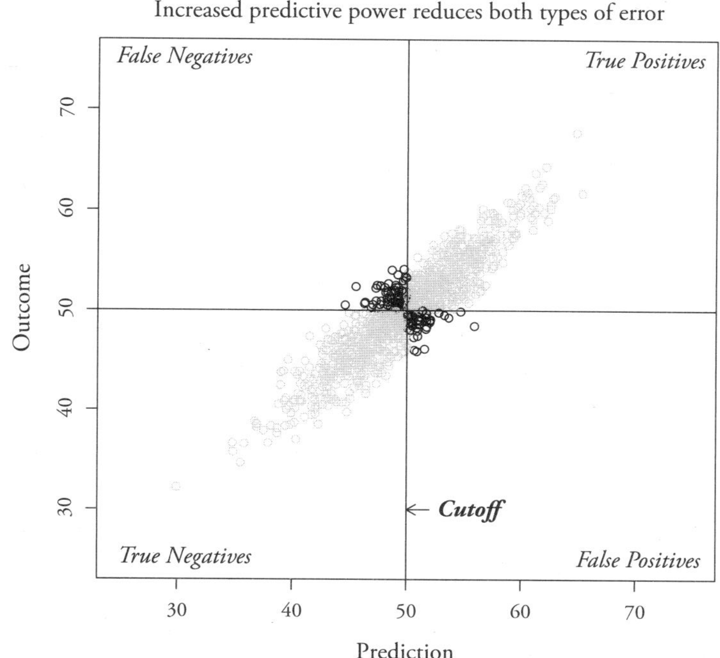

our cutoff scores, we marshal the resources to develop a new test that is a much better predictor of performance. Notice that the cloud of data is now narrower and more diagonal, and that both types of errors have been reduced considerably. So while scientific know-how can't eliminate trade-offs in the short run, it can sometimes greatly alleviate them in the long run, if we're willing to invest in the necessary research and development.[54]

EXAMPLE: DIAGNOSTIC TESTS

Another familiar domain where our values about the risks of various errors come into play involves diagnostic tests in medicine. We will use this example to illustrate another factor that influences error rates: the "base rate"—the percentage of cases where the signal is actually present.

A recent *New York Times* article documents that for several rare medical conditions (for example, DiGeorge syndrome or Wolf-Hirschhorn syndrome), almost everyone who tests positive will turn out actually to be negative—81 to 93 percent of the time for the diseases they examined.[55] Does this mean these tests are worthless? Fortunately, the vast majority of people taking the test do not have these rare conditions. The actual number of false positives we observe isn't just determined by the inaccuracy of the test (i.e., the test's false positive and false negative rates)—it is also determined by how common the disease is. When the test is imperfect *and* most people who take the test don't actually have the disease, we can find more false positives than true positives simply because there aren't that many true positives to find. (As you might expect, there is a way to calculate this; it's called Bayes Rule.[56])

Despite the high false-positive rates, it turns out to be worth it to test for these conditions so that the afflicted can receive the necessary treatment. The good news is that we don't usually leave the false-positive cases worrying for long or receiving unnecessary treatment. We often have a second test that is more expensive to administer but makes fewer false-positive errors. We can use the more expensive test on people who tested positive, correcting many of our errors.

FLYING (PARTIALLY) BLIND

We've been analyzing patterns of error by using 2 x 2 tables. But in a great many real-world settings, we never get to see all four cells of the table. Consider admissions decisions at UC Berkeley and Stanford, where we teach. We routinely (and accurately) tell our students that they are fantastic, the cream of the crop, and that we're lucky they joined us. Still, admissions committees know that if we had dipped farther down the list, there were a whole bunch of other fantastic people who didn't quite make the cut. But the thing is, we'll never know whether we made a mistake by not taking them, really, because we'll never see how they would have performed if we'd admitted them — or whether our admitted students might have thrived more at a competing university. We never get to see the universe in which we made the opposite decision.

An interesting case arose back in 1973. There is a long-standing debate about whether psychiatrists and parole boards are good at deciding whether a jail inmate or a resident in a psychiatric facility is too dangerous to be released. But because of a major budget crisis, some institutions had to let everybody go. So even though the psychiatrists had made judgments that some people were too dangerous to be released, the state officials said, "Well, we don't care what you said; we're going to let them go anyway." So we actually got to find out what happens when you release someone you consider dangerous. And it turned out that a majority of those judged "dangerous" did not commit any violent offenses in the next three years. It looks to us like the examiners were usually willing to accept a very high false-positive rate in order to avoid being responsible for letting someone go who turned out to be dangerous. They seemed to be privileging the safety of society over the rights of the individual. Readers will differ in their views about whether that balance is right or wrong.

THE HORNS OF A DILEMMA

We have seen that there is a painful trade-off between the risk of false positives and the risk of false negatives — it is hard to reduce one without increasing the other. Again, *setting a cutoff score is a policy decision* reflecting institutional decisions about the relative costs of the errors, and such decisions are inherently political decisions about values rather than scientific.

So let us reiterate the most essential point of this chapter. Science can tell us how to estimate probabilities, but it can't tell us what decision threshold we should use. Our standards of proof (our decision thresholds) are an expression of a value judgment: In a particular given situation, which error are we more eager to avoid?

Scientists and science enthusiasts sometimes overlook this point. Consider the debate about the advisability of lockdowns early in the Covid epidemic. Although we won't argue the point here, we think that there were compelling scientific reasons to believe that a lockdown could reduce virus transmission. But the risk of transmission must be balanced against any harms created by a lockdown. Eventually, we may be able to quantify each of these risks, but at the time the decision had to be made, we didn't know enough to do so. We think many people were correct to argue for the public health benefits of lockdowns given the state of medical science, but it does not follow that adopting lockdowns was what the science dictated, since lockdowns have many other kinds of effects. When we buy a car, each of us decides how to trade off the benefits of safety features against the cost of the vehicle. In the same way, there are various competing benefits and costs of mandating a lockdown, involving public health considerations but also economic, educational, and other dimensions. In chapters 16 and 17, we will return to the questions of how societies can work with the values that set their decision thresholds.

ERROR TRADE-OFFS AND "STATISTICAL SIGNIFICANCE"

Readers who've taken introductory statistics will recognize the same kinds of trade-offs in statistical hypothesis testing. You've heard the phrase "statistically significant," which is determined by a magical state called $p < .05$. That is the p-value — roughly, the probability of detecting an association as strong as the one you observed in your data, if in fact there's no association whatsoever.

This system is controversial. If you get a p-value that comes in at .049, then you go out and celebrate. If it comes in at .051, you're in despair. That strikes many people as irrational—to be so rigid about this tiny little difference in probability. Why do we get to call one significant and the other not?

Statisticians of an earlier generation, fairly arbitrarily, picked this .05 threshold. The decision has been made, by convention, to be more concerned with avoiding false positives than false negatives. But you could argue, on a principled basis, for higher or lower thresholds. Maybe science is harmed more when we say the effect wasn't there and thereby overlook a true effect.

It's troubling because in some areas of public policy, we test an intervention in the real world. The sample sizes are small. The measurement is noisy, and so it's very hard to detect the signal. And because we've set this very strict threshold that tries to keep people from claiming that an effect is there when it's not, we can't say there's an effect if the effect is real but small (because it is harder to achieve a p value $<.05$ in this circumstance). There's a good possibility that we've abandoned a lot of effective government interventions because we couldn't detect the effect, because we had to use this stringent threshold in domains where data was scarce and noisy. Deciding whether to cut these losses is a value judgment.

CAN WE MAKE TRADE-OFFS LESS EXCRUCIATING?

We've already seen that science can reduce decision errors by improving the quality of our prediction tools. And there are additional steps we can take to manage trade-offs.

Sometimes—and we would argue much of the time—we can adopt a course of action provisionally. We can try it, examine the results, and revisit our decision. Legislators can include "sunset clauses" where a program is evaluated before its funding is renewed. Admittedly, this does get tricky; sunset clauses create incentives for partisans to distort the evidence for or against a program.

And sometimes we can simply suspend judgment and collect more evidence before we make a decision. As we argue throughout this book, *sometimes it is premature to form an opinion.* Families defer decisions about home purchases or job changes when the economy seems uncertain. Doctors postpone intrusive procedures until test results become less ambiguous. Of course, in other situations we can't wait; we need a decision. For example, mayors need to decide whether to evacuate citizens when there is a flood warning; military officials need to decide whether to fire on a jet flying in restricted airspace. Jurors don't get to say, "Well, we think it's 82 percent likely that he is guilty, Your Honor." If they did, the judge would probably send them back to the jury room until they made a decision.

Our hope is that this chapter has helped the reader to appreciate that public disagreements aren't always about *the facts.* Sometimes they are disagreements about setting the right *standard of proof*—the amount of evidence that we will consider strong enough for one side to change their mind.

This is an inherent feature of probabilistic thinking—we can't expect certainty, so we all need to decide how much evidence is enough. And because of the lack of absolute certainty, even the best evidence-based policy will make errors. You and I may agree that it would be bad to fail to help people in need, and that it would be bad to let other people exploit the government's generosity. We might just disagree about which error is worse. Articulating our views about such issues won't eliminate disagreements, but it can greatly clarify them. And both sides can agree that improving the accuracy of our forecasts will reduce both types of error.

CHAPTER 9

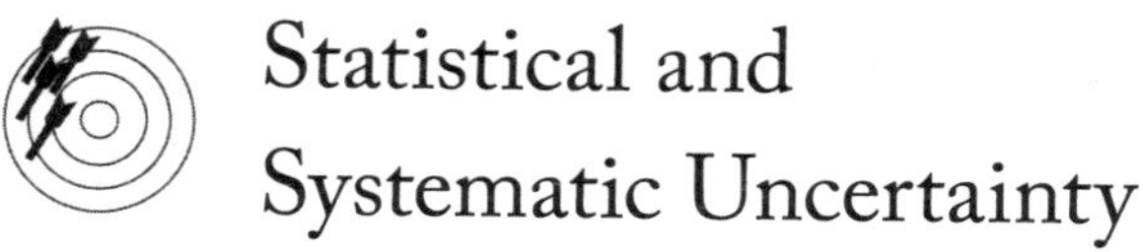

Statistical and Systematic Uncertainty

If only the objects in the world and their properties would be just one thing or another! If everybody was either exactly three feet tall or exactly five and a half feet tall, then it would be easy to tell if your five-year-old was tall enough to get into most of the rides at Disneyland before you bought tickets online and drove all the way there. Instead, you have to measure your enthusiastic offspring, who may in fact have reached the forty-inch requirement, and all you have is an old measuring tape that you found in the back of a drawer. It turns out that your child's height is quite close to the "tall enough to ride" requirement, and as your hand jitters holding up the tape measure—introducing noise into the measurement—you think, Yes, and then you think, No. You start to think that you might just make a whole bunch of measurement attempts and then average them—that should handle the hand-jitter noise. But then you vaguely remember having left the tape measure in your pants pocket while it went through the wash, and you wonder if it might have shrunk. (The pants sure did!) If so, averaging multiple attempted measurements won't help. Sigh. . . .

This poignant story introduces one further conceptual tool useful when we are faced with noise and uncertainty in the real world. First, there are two ways in which noise and uncertainty can affect measurements: sometimes measurements are randomly bouncing around but can be averaged out (like your unsteady hand holding the measuring tape);

other times they are pushed in one systematic direction or another (*every* measurement comes out overestimated if you use a tape measure that shrank in the wash). If we're really going to try to quantify our credence levels and recognize the odds of random numbers giving us false impressions of the world, we have to be able to handle both sorts of complications.

Science offers a useful language and approach for this purpose. In fact, knowing how best to handle these different varieties of noise and uncertainty ends up being such an important problem that several fields of science have developed terminology for it. The terms differ, but their meanings are similar. Physicists like Saul typically call the two kinds of sources of noise *statistical uncertainty* and *systematic uncertainty*, while social psychologists like Rob might talk about *reliability* and *validity*. Statisticians might differentiate between precision and accuracy, or they might use the more technical-sounding terms *variance* and *bias*. (This is a technical use of the word "bias" and should not be confused with the ordinary meaning.) Since all these terms have slightly different nuances (we will come back to some of these later in the chapter), John, the philosopher, gets to choose whichever fits his meaning best.

THE BATHROOM SCALE

To understand these terms, let's begin with another concrete example that illustrates how these two different types of uncertainty show up in day-to-day life. Imagine that your doctor has advised you to lose five pounds for health reasons. Right afterward, you head off on a business trip, and at each hotel you check your weight. Every day, in every hotel room, you are using a different scale. The first time you weigh yourself you say, "Oh, that's funny. It looks like I'm three pounds lighter than I thought I was. Maybe the scale's off." Then you travel to the next town, try the next bathroom scale, and say, "Oh, look. This one says I'm about two pounds heavier than I thought I was." You start to suspect that hotels don't bother calibrating their bathroom scales that well. Over the course of a whole month of traveling, you visit lots of bathrooms and lots of scales, and you figure that if you take an average of all the measurements, it will probably be about right, because the odds that every scale is off in the same direction seem slim to you.

When you return home, you use your own bathroom scale. What you don't know is that your scale at home is biased—that it always gives a reading five pounds lighter than it should. Maybe at first you think your weight is low because you are dehydrated from the flight, but after a month of weighing yourself every day with *this* scale, you are still being told you are five pounds lighter than your average weight when you were traveling. In fact, it doesn't matter how many times you weigh yourself with this scale; you always congratulate yourself on your consistency in keeping off those five pounds. Of course, that's exactly the kind of situation that scientists are constantly trying to invent ways to recognize: you want to be able to tell whether or not you're in a situation in which you got a very nice reproducible result—that is, you weighed yourself many times and always got the same number—but it is five pounds off from the truth.

This is where scientists start to differentiate—and label—the two kinds of uncertainty: *Statistical uncertainty* describes those sources of noise that result in your measurements randomly scattering around the correct value, some higher and some lower, like the varying weights on the hotel scales. You can always get closer to the true answer by averaging more and more measurements when they only have statistical uncertainties. *Systematic uncertainty*, on the other hand, describes sources of noise that push every measurement in a single direction, either all high or all low, as with your inaccurate home scale that pushed every measurement low. It doesn't matter how many measurements you make when they only have systematic uncertainties; they won't average to the true answer—they will average down to a "biased" result.[57]

The interesting thing about these two types of uncertainty is that once you are aware of the difference, it becomes clear that to make a good measurement (that is, with a clear signal, not too confused by noise), you need different strategies for handling them. If your measurement keeps bouncing around randomly every time you repeat it (maybe your hand is shaking slightly as you try to hold the ruler very steady), you will need to take many more measurements so you can average down this statistical noise—or find somebody with a steadier hand, who can get more consistent results. When you suspect a possible *systematic* uncertainty, that is generally much more challenging, so much of a scientist's effort—and much of the rest of this discussion—centers on ways to address it.

HANDLING SYSTEMATIC UNCERTAINTY

When addressing possible systematic uncertainties, you get a chance to be creative, because your first job is to try to propose all the possible ways that the measurement could be systematically pushed in one direction or the other. Is it possible that the tape measure got stretched or shrunk, so that every measurement you make is off in the same way? Do most people, perhaps unintentionally, shop around for bathroom scales that underestimate their weight, so that most scales you come across at a friend's house will underestimate? Your second job is to invent ways to check for this systematic bias, or, better still, limit its effect on the measurement. Can you find other tape measures, or bathroom scales, that you trust more to compare lengths or weights? Alternatively, whether or not you can prove there is a systematic bias in the measurement, can you find a way to make the measurement so that it won't matter if there is a specific bias in your measurement tool? This last possibility sounds counterintuitive—how could a specific bias be irrelevant?—so we should look at an example.

Imagine that Sarah is preparing to run a mile, which is a four-lap race around the track. Her coach wants to time how fast she runs the third lap of the four-lap course. So every time Sarah goes past the line on the track that marks the start of the third lap, her coach presses a button on the stopwatch to start a timer. When Sarah goes past that line again (starting the fourth lap), her coach presses the button again to stop the stopwatch.

As we start to examine this timing measurement, we first think of several potential sources of statistical uncertainty: Sarah may not run her third lap at exactly the same speed every time, and Sarah's coach may not be consistent at pushing the stopwatch button at exactly the same moment every time; sometimes he may be a little fast on the draw and sometimes a little slow. These are problems we can handle by averaging the same measurement taken at different training sessions (presumably not all on the same day, since Sarah might get tired). In fact, in many circumstances, statistical uncertainty is reduced by averaging across the individual measurements. But what if the coach consistently has a delayed reaction to Sarah crossing that line on the track, so that he is *always* a little late pushing the stopwatch button? This sure sounds like a source of systematic error: that delayed reaction won't get averaged away with more measurements.

You may have noticed, however, that the way this story was told, the measurement involves the coach pushing the same button with the same delayed reaction twice, once to start the timed lap and once to stop it. So as long as the coach's delayed reaction at the start of the lap is the same as it is at the end of the lap (which is also the start of the following lap), the two will cancel each other out and the measurement will not be biased. This may sound like an artificial example where we just got lucky, but this is exactly the sort of creative solution to an annoying possible source of systematic uncertainty that scientists always look for. Remember how useful it was to randomly assign people in a medical experiment to the group who got the drug or to the control group who didn't? This was a clever way to ensure that sources of systematic uncertainty that might bias the result in one way or the other would get canceled out when the two groups were compared.

REAL-WORLD SYSTEMATIC UNCERTAINTIES IN ACTION

Our goal here is to be sensitized to these sources of systematic uncertainty—and such possible fixes for them—wherever they arise. Let's consider two real-world examples that are dramatic in different ways. The first is a practically important example of systematic uncertainty that you might have experienced if you've voted in local elections for more obscure candidates like members of the school board, and you hadn't been following the issues or the candidates for that office.

Studies have shown that when voters are presented with people's names on a ballot, and they don't know anything about the candidates, then, all else being equal, people will vote slightly more often for whoever is listed first on the ballot. The candidate listed first gets about 5 percent more votes than anyone farther down on the ballot, so this is a really dramatic effect because in many races that 5 percent is easily enough to win.

Perhaps surprisingly, you get the exact opposite effect if you do this verbally, not visually, which is what happens when pollsters call people up and ask them who they are most likely to vote for. In that situation, when the list of candidates is conveyed, the *last* person to be listed gets a boost. So if you think about it, if there happened to be an election with a race in which one candidate's name was printed at the top on the ballot and the pollsters read the list of candidates to people over the phone in ballot order, the preelection polls could get the predicted winner completely wrong: the top-of-the-ballot candidate would have a five-point disadvantage in the telephone polls but a five-point advantage in the election.

This is crazy! Knowing this source of systematic bias, why would you ever print all the ballots with the candidates listed in the same order? In California, in races for state legislative offices, somebody realized the problem of systematic bias in the order of the names, but then tried to solve the problem by randomly choosing an order for the candidates on the ballot and then printing *all* the ballots in that order. Clearly this doesn't solve the problem. You might say it's fair if you have a funny kind of idea of what counts as fair, because every candidate has an equal chance at winning the 5 percent boost by being put at the top of the ballot, but obviously that's not what we're after. We're not trying just to be fair in our mismeasurement; we're trying to capture what or whom those people who do have a preference would want to vote for, undistorted by the 5 percent of people who apparently choose based on name order. Oddly enough, for California statewide and US congressional offices, they don't make the same mistake: ballots are printed for distribution county by county, and the order in which candidates are listed on those ballots rotates with each of California's 58 counties. While it isn't perfect, this approach should help make the first-name advantage average out—just as your weight measurements average out on your multihotel business trip.

Our second real-life example illuminates how high the stakes can be when it comes to recognizing and addressing both statistical and systematic uncertainties. In today's world, whether our measurement of the global temperature change over the past century is wrong or right is crucially important. If we get our understanding wrong (our probabilistic understanding, of course!) we could cause great suffering, either by acting incorrectly to address a misunderstood problem, or by not acting and allowing a climate disaster.

During the past century, meteorologists around the world have been keeping track of their local temperatures using thermometers that they check each day. Climate scientists have used those measurements to help estimate the change in average global temperature, leading to the graphs we showed in chapter 6 that indicate that the temperatures have risen over the past century. Here's a situation where both statistical and systematic uncertainties are clearly in play. First, there's all sorts of noise in the daily local measurements, due to minor weather differences place to place, day to day, and year to year, and due to badly built thermometers. These sorts of uncertainties need to be identified and understood as "statistical uncertainties," since, with sufficient data, they could all be averaged away. Much more dangerous would be any sources of systematic bias that were different toward the end of the twentieth century than they were at the beginning.

Systematic errors that stayed the same over that time would cancel out (like Sarah's coach's delayed reaction time), so we don't care about them. For example, if everybody all over the world measured the average daily temperature only in the height of the afternoon, this would be systematically biased as warm compared to the average of the day. However, as long as this mistake was made throughout the century, it wouldn't contribute to a systematic bias in the measurement of temperature change.

On the other hand, if a measurement bias does change over the century, then we have a systematic error to consider in our measurement of temperature change. Consider the example of systematic changes in *where* the temperatures were recorded. During the early part of the century, more of the measurements would have come from Europe and North America, where the recorders were; but over the course of the century, Africa, South America, and Asia would have contributed more equally. Another geographic change over time could have come from urbanization. As the century progressed, more and more of the temperatures were recorded in or near cities, because more and more people lived in cities. And we know that cities tend to be warmer than the surrounding rural areas because of what we call the urban heat island effect.

These are exactly the sorts of systematic uncertainty concerns that have been raised in recent decades as the question of global warming has raised important policy questions. One of the groups that addressed this concern did a controlled study in which they compared temperature changes over the century in places that were close to cities versus ones that were in rural areas. (Saul helped with this project.) This approach made it possible to effectively control for this source of systematic uncertainty. In this case, they found that the difference *between* urban and rural temperature changes was much smaller than the warming effect we are concerned about, indicating that urbanization over the century was not a key systematic bias on the measurement of the global temperature change over the century.

What about the time of the day you take the temperature? It does turn out that there was a shift in the practice of how and when temperature readings were taken for use in calculating daily averages. For example, at the beginning of the century, the US weather bureau recommended that temperature readings be made around sunset, whereas by the end of the century, most US measurements were made in the morning (when it is probably cooler). So now what counts as the average temperature during each period is harder to establish because you have to try to correct for that difference in practice.

Another change in the practice of making temperature readings occurred as well—this one involving the temperature of the ocean's surface. In the early part of the century, they measured the temperature of the ocean by throwing a bucket overboard from a ship, pulling the bucket in, and measuring the temperature of the water in the bucket. Around the Second World War, they switched to putting a thermometer in the intake port that feeds water to cool the engines. As you can imagine, the two methods gave slightly different results. For one thing, the water going into the intake port is farther below the surface than the water hauled up in a bucket. And apparently, it never occurred to the people doing the measuring that anybody was ever going to want to compare the temperature records, so they didn't cross-calibrate. They were taking temperatures for other purposes. So now in all the studies that are done using ocean temperatures, scientists need to allow for an offset of unknown size between the ocean temperature measurements taken with these two methods. And this offset is hard to determine today because you can't get ships like those at the time of the switch in measurement method and compare the bucket measurement to the intake valve measurement.

Once we know about such changes in methods for collecting temperature data, those changes can become the object of study. In the examples above, the changes produced "temperature measurement offsets" that we need to measure, with their own statistical uncertainty. They certainly contribute uncertainty to the final best estimate of the global temperature change, but the good news is that once they become just another quantity to measure along the way to the answer, we can use the standard probabilistic reasoning that we described in previous chapters to quantify the uncertainty and to choose thresholds for what counts as "certain enough" to take action. For example, we could decide that the consequences of a global temperature change are major enough that we will act even if we have just 75 percent confidence in a rising temperature, or we could decide that the shift in resources is disruptive enough to a global economy that we need 95 percent confidence.

THE CREATIVE CHALLENGE OF SYSTEMATIC UNCERTAINTIES

The example of identifying many possible sources of systematic bias in the measurement of global temperature change demonstrates that there are different routes to dealing with each of them: in some cases we find ways to cancel them out in the measurement; in others we measure the systematic effect, turning it into a more manageable statistical uncertainty; and in yet other cases (not given here), we simply need to show that the size of the systematic effect can't be so large as to matter to our measurement. The bottom line is that *once we have identified* these possible sources of systematic uncertainty, they can be studied and accounted for. A huge amount of the training of scientists therefore involves just getting better and better at recognizing such sources of uncertainty, and then thinking up creative ways to control, balance out, or measure each one well enough so that it doesn't make the measurement too uncertain to base a decision on.

In later chapters, we will be discussing the importance of working with people who don't agree with one another; one of the most important reasons being that it is dramatically easier to uncover sources of systematic uncertainty when you are challenging someone else's position! So if you are going to find systematic uncertainties that could be distorting your measurement, you probably want to reach out to people who will challenge that measurement (as painful as that may be). At its best, the tradition of scientists reviewing and strongly critiquing one another's work is an important further element of scientists' hunt for systematic uncertainties, even beyond all the training of scientists to find them themselves. Once you become attuned to the problem of systematic bias, you are likely to start seeing, or looking for, systematic biases in the evidence for decisions you make in your own life. (For example, when your boss critically edits something you write or disagrees with something you suggest, you are quite aware of it, but not so much when they just accept your work; this systematic bias in your self-evaluation evidence might incorrectly lead you to look for another job.) However, outside the culture of science, it's harder to get yourself to find someone who disagrees with you and can help hunt for your own systematic biases. (In this case, the term "biases" can include *both* the more technical meaning we have been discussing *and* its usual definition.)

Moreover, when we need to make our own decisions based on scientists' results—which medicine to take, say, or whether to vote for a policy on fracking for natural gas—we should now be sensitized to check whether the scientists have explored a good range of systematic uncertainties in their research. For that matter, we might want to know which contrarian or rival scientists have already looked for systematic uncertainties. And these are the sorts of questions that we should expect our experts in a field to address when they explain why they believe a scientific finding.

A MNEMONIC ICON

There's an iconic visual that is often used to make clear and memorable this chapter's distinction between statistical uncertainty and systematic uncertainty. If we imagine that our attempts at measuring some quantity are like a game of darts, where each dart is an attempted measurement, then we can think of the spread of darts around the bull's-eye on the dartboard as representing our uncertainty in the measurement—by how much we are missing the mark of the true answer. As we now know, our measurements can miss hitting the true answer in two different ways, either by being randomly spread around the correct result (statistical uncertainty) or by being shifted in one direction from the correct result (systematic offset). With the darts, we would represent the former, statistical uncertainty, with a cloud of darts around the bull's-eye, and the latter, systematic uncertainty, with darts all hitting to one side of the bull's-eye. More statistical uncertainty means that the cloud spreads farther out around the bull's-eye, and more systematic uncertainty means that they shift farther away from the bull's-eye. Of course, usually our measurements miss in both ways at the same time, so our cloud of darts is spread out (systematic uncertainty) *and* its center is offset from the bull's-eye (statistical uncertainty). The dartboard icon below shows a representation of four possible outcomes of a series of measurements, considering the better-or-worse options for both of these sources of uncertainty.[58]

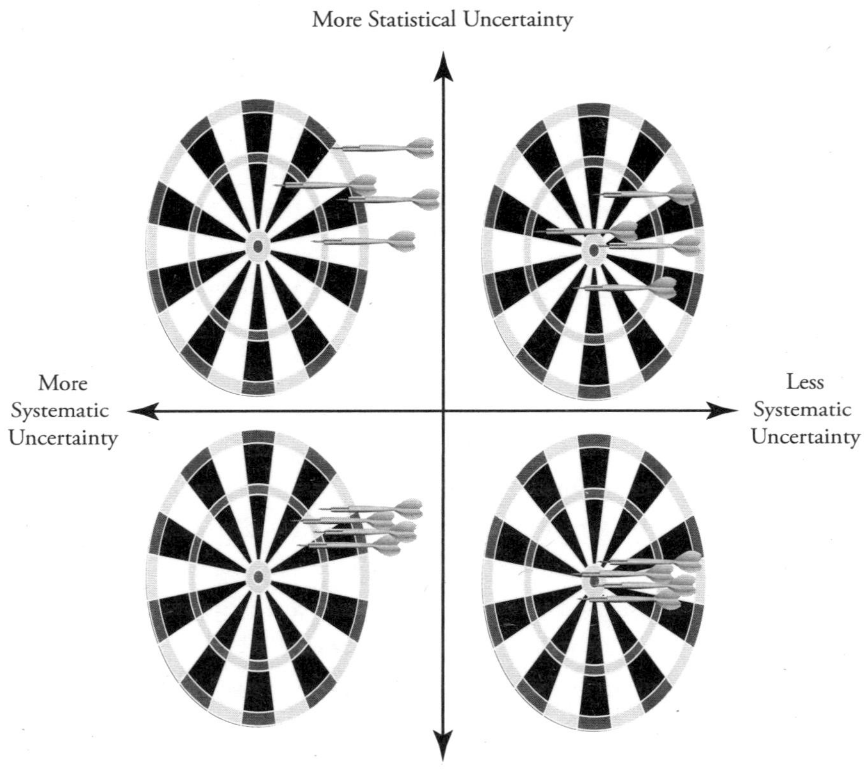

This is perhaps one of the more concrete ways to understand the distinction between the two kinds of uncertainty—and it may also help explain what the other fields' terminology is referring to. For example, you might expect the tighter clustering of the darts on the bottom dartboards to be described as "less variance," while the better centering on the bull's-eye on the right-hand-side dartboards would be considered "less bias."[59]

TRIANGULATION, REDUX

Given that the name of the game is systematic uncertainty, versus the easier-to-handle statistical noise, you might start to feel worried at this point. This systematic uncertainty stuff sounds hard! Not only do you have to be creative to think up the ways it can fool you, but you can't guarantee that you have identified all of those ways! And we haven't even discussed the systematic errors that enter the story that are due to real human bias—what we discuss in chapter 12 as "motivated reasoning."[60]

Before we really start to panic (how can we ever trust any decision based on a measurement?!), let's look at some elements of scientific thinking that can not only keep us calm but give us a fighting chance to build confidence in our handling of these uncertainties. The first of these is the use of triangulation.

Back in chapter 2, we explained triangulation as the use of a variety of instruments—sometimes enhancing a variety of our senses—to home in on a better picture of reality than you could get with just one instrument. For our current discussion, this kind of triangulation brings an additional advantage: If we use sufficiently different measurement approaches, they will each have different sources of systematic uncertainty. If we then find the same result with these multiple methodologies, it becomes much less likely that these different systematic uncertainties have all "conspired" to give the same distorted (incorrect) results.

For example, suppose we were trying to measure the current voting intentions of the eligible citizens in a certain city for an upcoming election for mayor. Imagine that we did a good job of reaching a large enough random sample of voters that we were pretty sure that our statistical uncertainty was low.[61] Now let's say that we were particularly worried about the two sources of systematic uncertainty that we discussed earlier—the biases toward the first candidates on a written list and toward the last candidate on a spoken list. If we can get half of our survey to be done verbally on the phone and half done in writing on the internet, then we can triangulate between these two systematic uncertainties, since the two different polling methodologies each introduce a different one of these systematic uncertainties. If the two types of polls agree, then we can be pretty sure that neither of these systematic uncertainties is important; and if not, we can use the extent to which they differ to estimate the size range of any bias due to one or the other (or both) of these systematic effects.

Now, having drilled down to this important tool of scientific thinking, and having focused on the statistical and systematic sources of uncertainty, let's zoom back out to the big picture. We have needed to explore these sources of uncertainty so that we can better manage them in our decision-making, when our decision is based on measurements intended to help us track our shared reality. In part II of the book, we have seen how this "reality-based" decision-making relies on our ability to understand reality probabilistically. In part I, we also described the techniques to identify the "causal levers" we use to solve problems and change our world—and we should remind ourselves that these techniques, too, result in probabilistic confidence in the causal levers, with all of the same need to identify statistical and systematic noise that might degrade this confidence.

This sounds like a lot to keep track of—and it is! Next, in part III, we turn to another of the secret weapons of science, the one that allows us to keep our heads amid all the causal factors, probabilities, sources of systematic uncertainties, and detection thresholds. It begins with something we might call the can-do aspect of scientific thinking.

Part III

The Radical Can-Do Stance

CHAPTER 10

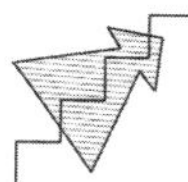

Scientific Optimism

What is the longest time you've ever spent trying to solve an intellectually challenging problem or a puzzle? Is it ten minutes, two hours, a day, a month, a year, a decade? Most people we have asked have trouble thinking of examples of problems and puzzles they've spent more than a few hours on; perhaps, at most, a few days. But how many tough problems in the world can be solved in less than a few days? The world isn't that easy! For that matter, less than a month doesn't really do justice to a problem worthy of the readers of this book.

Sticking to a problem long enough to solve it is a challenge that comes in all sizes. How long are you going to keep trying to land a man on the moon if you are leading NASA in the 1960s? How long are you and your spouse going to keep trying to put together that IKEA cabinet for your study? (There aren't *that* many ways to misinterpret the illustrations.)

The thing is, we humans are inherently lazy. It's not our fault; we probably evolved that way to conserve our energy. And, oddly enough, it *feels* like we use a lot of energy to think hard,[62] so we avoid it when possible, just like we avoid walking up a steep hill if we can go around it. But we generally need to think hard to solve important problems. Among other demands on our slothful brains, we have seen that it takes some appreciable mental work to recognize when we are likely to fool ourselves by mistaking a spurious pattern for a signal in noise, or to construct a list of possible sources of systematic uncertainty that will bias a crucial measurement. And in part IV, we will discuss even more such mental challenges to thinking clearly.

To make things worse, not only are we lazy, but even one of our positive attributes, our wonderful curiosity about new things, exacerbates matters. After a day or so a problem doesn't seem that new, and we're inclined to move on to being curious about something else. In addition, while we apparently really like our curiosity to be rewarded by obtaining a new understanding of the world, and our desire for that reward is a good incentive for us to focus on a problem, we also get frustrated when we don't make quick progress and get this reward after a minimum of effort—an unfortunate intersection of our curiosity with our laziness![63]

So what are we going to do about this very human inability to stick to a problem? This is where a rarely discussed, hidden tool of science comes into play. It consists of a simple mental trick that the culture of science invented along the way that we will call *scientific optimism*. This isn't your everyday, garden-variety optimism. Scientific optimism is basically a can-do spirit, the expectation that the problem at hand is going to be solvable—by you, or by you and your team. When a complicated problem comes along, you're much more likely to solve it if you approach it as if that solution is within your grasp. Essentially, scientists have invented ways to fool themselves into believing that they can solve a problem, just long enough to actually do so—the one time in this book that the goal isn't to avoid fooling yourself!

There are many examples in history where, apparently, people thought something was not doable, but then a rumor went out that somebody somewhere in the world had figured it out and suddenly lots of people were able to do it. They would start saying, "Wait, those folks know how to do it, it's got to be doable," and they would just keep trying. "Oh, maybe they did it this way.... No, that didn't work. Maybe it was this other approach...." Once they saw that it was possible to solve the problem, they found the motivation not to give up. And in the end, they may have figured out a completely different way of solving the problem than those who first solved it did.

You can think of this as the mental equivalent of all those athletic records that people thought would never be broken, like the four-minute mile. We have all heard the stories of how these seemingly unbeatable human limits were almost routinely surpassed once one person finally showed it was possible. Translating this to our more cognitive problem-solving situations, simply imagine the difference between trying gamely to build a cabinet out of a box full of parts from various bits of discarded IKEA cabinets, versus building one from a new IKEA cabinet kit that several of your friends also bought and told you they had successfully assembled. Certainly, the second situation would keep you going much longer than the first. But the extra twist in scientific optimism is that we temporarily adopt the belief that we can succeed even when we *don't* know that this is a complete IKEA cabinet kit that works, and in this way we give ourselves enough time to tackle a difficult problem. The scientists need scientific optimism because they are trying to reach new discoveries, but we all need it because we often have to approach problems without a guaranteed solution. (A dramatic contrast to scientific optimism is the phenomenon of "learned helplessness." Apparently, both humans and other animals will give up on trying to resolve an uncomfortable or painful situation if they have too much experience of such situations being outside their control. Even if they could actually remedy the current situation, they have learned not to even try.)

One of the most amazing examples that we know of, of how thinking that a problem is solvable made it possible to solve, is Fermat's last theorem. In 1637, mathematician Pierre de Fermat wrote this in the margin of a book:

> It is impossible to separate a cube into two cubes, or a fourth power into two fourth powers, or in general, any power higher than the second, into two like powers. I have discovered a truly marvelous proof of this, which this margin is too narrow to contain.

For the next 358 years, mathematicians worked on this problem because they all thought, based on Fermat's claim, that it was solvable. And in 1995, the problem was solved. Probably not at all the way Fermat did it, but it was Fermat's confident claim that it was doable that kept people working on the problem—*for 358 years!*

Saul first started to notice the importance of scientific optimism as a graduate student. He was trying to choose a research group and came across one that had an amazing can-do spirit that the professor leading the group, Richard Muller, had instilled. Any exciting project was fair game in Professor Muller's group. There was a sense that if you needed a new tool, you could simply invent it. If you needed to build something, you built it. If you needed to learn a new area, a new field, you learned it, whether complex electronics or DNA manipulation techniques. This can-do spirit inspired the group to enthusiastically take on a wide variety of problems and challenges. At that time, the group was working on a technique to measure the deflection of light bent by the gravity of Jupiter, inventing a mini tabletop cyclotron, measuring carbon in the atmosphere above the oceans to help understand the earth's carbon cycle, and developing the first automated telescope system to discover relatively "nearby" supernovae. This scientific tradition of being comfortable taking on whatever challenges a compelling problem presents is perhaps one of the greatest strengths that science offers us, whatever our walk of life.[64]

Saul began his graduate student research with that last project, the automated supernova search. Eventually that project led to an even more challenging one when it became clear that the same techniques could be used to find vastly more distant supernovae that could reveal the history of the universe's expansion—and predict its eventual fate. This was going to be an immensely challenging project. Saul and his team estimated it would take three years to find the dozens of such distant supernovae they needed to find in order to measure changes in the universe's expansion rate. Three years later, zero of these distant supernovae had been found. (Tip: When at all possible, avoid fields where your success depends on clear weather!) After five years, the first one was discovered. After seven years, the team had really figured out how to do the project, and they were finding distant supernovae in batches of half a dozen or so. Nine years in, they had the dataset in hand, but still hadn't figured out how to analyze it at the level required. After ten years, they had an answer—and a big surprise: the universe's expansion is accelerating.

Interestingly, at every stage of the ten-year supernova project, the team was pretty sure that the project was doable. Scientific optimism did keep them going—but it is important to see how this works in practice. At every step of the project, the team could see what successes had been accomplished and what was needed next to keep moving toward the goal. This iterative advancement is key to so much of this can-do story; it's not that the scientists are counting on accomplishing their goal in one fell swoop.[65]

Making progress on hard problems, problems that can take months, years, or decades, usually requires such iterative advancement, meaning we do better and better with each attempt and build on what we learned in the previous attempts. As we unpack some of the practical implications of scientific optimism, this concept of iterative advancement is one that should resonate with all planners. To illustrate with just one example: policymakers and legislators drafting a major welfare reform bill or education improvement bill or crime reduction bill could—and should—build into it the assumption that it will need to be iteratively improved every few years as we learn what works and what doesn't. Such policy-updating mechanisms are not unheard of, but they are certainly not a prominent, visible feature of our government programs, at least in the United States. Perhaps if they were, we would have more of a sense of the degree of progress on any of these societal goals.

MORE PIE

The day-to-day can-do spirit with which a scientist will (often) approach a new problem has a rather different practical implication when it is applied to problems of shared resources. Many social conflicts arise from perceived scarcity: different people or groups make a claim for a resource and there does not seem to be enough of it for all of them to get what they want. (This is sometimes called a zero-sum game because whatever I get, you lose, and vice versa.) For example, should limited water resources go first to rural agriculture, or to urban developments? Should a polluting coal-fired power plant that generates electricity for the entire city be situated nearer to one neighborhood or another? Such decisions can seem to be intractable sources of conflict between well-meaning people who hold strongly divergent philosophical views on the principles upon which we should build our societies—and, of course, the conflicts can bring out the worst of our natural competitive tendencies to take the biggest slice of pie that we can, or to vanquish our foes.

But note that we said "*perceived* scarcity." The claim that something is scarce is usually based on some underlying assumptions that might be wrong—a failure of imagination. And that's where the can-do attitude that comes with "scientific optimism" can be helpful.

What scientific optimism's stick-to-itiveness can bring to the table here is the alternative possibility of enlarging that pie so that we are not forced into a decision where one person's gain is another's loss. Today the world's population is vastly (four times) larger than it was 100 years ago, yet the fraction of people living in extreme poverty has dropped from almost 60 percent to under 10 percent. This means that the absolute number of people living in extreme poverty has declined even though there are some 6 billion more people. Clearly, the diminution of extreme poverty was not primarily a consequence of taking away resources from one group and giving it to others, but rather of a dramatic increase in the production of resources. Now we are concerned about the group still living in poverty, and about the effects on the environment of global production growth. Again we will need every bit of our iterative problem-solving abilities together with the scientific can-do spirit.

Another interesting example of growing-the-pie solutions can be seen in the efforts to cut carbon emissions caused by energy production and consumption. In recent years, new energy technologies that produce less greenhouse gas (wind, solar, geo, hydro) have been developed and existing ones improved, to the point that some experts believe that these technologies are now cheaper and safer to use than the older energy sources that emit more greenhouse gasses. If so, political fights over who gets to use which energy technologies become moot. Despite differences in people's degree of concern about greenhouse gases, if it's cheaper not to produce them, then everybody is on the same page.

The history of science is so replete with such stories of leapfrogging a stumbling block that it can provide us with a different approach to a tough negotiation—the possibility of making everyone in the room come out ahead. This is particularly important to recognize in a world saturated with media that often appear to have business plans based on scaring us with everything that goes, or might go, wrong in the world. The natural reaction is to hunker down and try to protect everything we have, generally making it extremely difficult, if not impossible, to find the grow-the-pie win-win solutions. Scientific optimism as a cultural antidote to media's doomscrolling offers a different starting point.

POLLYANNA, SHMOLLYANNA

With all this optimism in the air, what happened to our book's running theme about science hunting for all the ways that we fool ourselves, and developing techniques to avoid them? It's all very well to have a can-do spirit and to look for iterative progress while we bake bigger and better pies, but there is such a thing as beating your head against the wall. Sometimes a problem just isn't ripe to be solved; you have to know when to call it a day and try working on a different problem. Although we talk blithely about *fooling ourselves into thinking we can solve a problem just long enough to actually do so,* we certainly don't want to fool ourselves longer than that!

The point we want to call attention to is that we are generally far too biased toward giving up too early, and that we need scientific optimism to keep us alert, alive, and motivated despite the inevitable setbacks in any challenging work. But along with this cultural mental trick of scientific optimism, science also provides us with some very practical tools to help us take on these problems that might seem overwhelmingly big, tools that can help us parse these problems into more manageable pieces and perhaps help recognize when the next step really is too big. We will discuss a few of these tools in the next chapter.

Even this current chapter's discussion already gives one interesting clue: If we find ourselves sticking to a problem simply because we have already invested much effort in it, this may be a warning indicator—and it may be time to quit. The time and resources already sunk into a project do not by themselves provide a reason to continue. (As some readers will know, there is even a name for falling into that mental trap, the "sunk-cost fallacy.") On the other hand, if we are seeing incremental but real progress, this may suggest that we are still on the path toward a solution through iterative improvement, and we should deploy our best scientific optimism and stick with it.

And even when we conclude that there is not enough iterative progress and we should hold off on our relentless "scientific optimistic" pursuit, this is sometimes more of a pause than an abandonment of the goal. Sometimes the various pieces of a solution that need to come together are not all ready at the same time and we need to set the problem aside until, for example, some new supporting technology is available. In fact, a big part of a scientist's can-do spirit can come from the ability to remember a long-standing problem and recognize, when some technology falls into place, that it can now be solved. (The solution of Fermat's last theorem had this character in that results in the 1980s in fields of mathematics that hadn't existed in Fermat's time opened up the possibility of the 1995 proof.)

Of all the 3MT concepts and topics discussed in this book, scientific optimism by its very name is bound to be the most upbeat. So many of the other topics deal with what you have to be concerned about to keep from fooling yourself; how to apply the crucial brakes on our natural self-delusions. But you can't drive a car with brakes alone! Scientific optimism is the necessary accelerator pedal that keeps us moving and, ideally, making progress. True, you very often have no guarantee that you will solve the problem. You are not going to be able to get closure at any given moment. You're instead constantly playing an improvement game any given day, month, year, or decade. That can sound unsatisfying, but if you get the right mental take on it, it is one of the best pleasures in life—to feel that you are getting there through iterative stick-to-itiveness.

With these thoughts about the culture of scientific optimism bouncing around in our heads, we start to notice that there is often an opposing culture—not the culture of healthy skepticism, those necessary brakes on self-delusion, but rather one of fashionable cynicism. We have probably all felt the pull of appearing smart by showing that we have a "world-weary wisdom": we've seen it all before and we know that this hopeful effort, whatever it is, is doomed to fail. "Like that'll ever work" is a great conversation stopper—and a cynical comment can often end what could've been a productive conversation. Part of our job is to catch this cynical culture when we see it—in ourselves, in others, and in the media—and allow our can-do scientific optimism to play its role in balancing our more measured healthy skepticism so we can sometimes, maybe often, solve problems and expand the pie.

CHAPTER 11

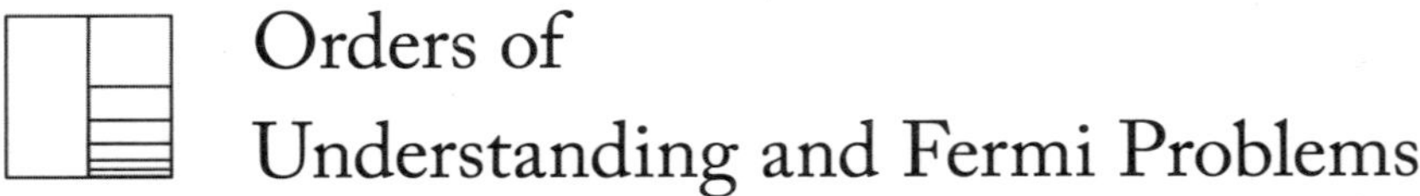

Orders of Understanding and Fermi Problems

Let's say you are now "cooking with gas," ready to use all your best scientific optimism to keep going on a big, complex problem.[66] How do you even begin to approach such a problem? And how do you check apparent results along the way? The world, of course, is *much* more complicated than even the busier diagrams in chapter 3 illustrating how many possible factors could be causally affecting a measurement we make or an outcome we observe. You might start to worry that you're going to be in real trouble if you have to express your degree of confidence in almost any causal relationship in order to decide what you're going to do about something. It seems questionable whether you could ever get above a confidence level of 55 percent. Real-world problems—like finding the "levers of change" in complex situations—require us to use more conceptual tools from the science toolbox. We'll begin with one that we call *orders of understanding.*

The world *is* a pretty complicated place. In almost any real-world issue, you have to take into account lots and lots of factors. The problem is that we're not that good at holding lots of complexity in our head at one time. You may have heard that humans can hold only about seven items in their short-term memory at any given moment. It turns out that there are all sorts of caveats to this particular claim, but you probably have at least an intuitive sense that you are not very good at holding, and working with, more than just a few ideas in your head at the same time.

The idea that scientific thinking brings to the table is that not every one of the many causal factors in a real-world problem is equally important. You can figure out most of what is happening—and what *can* happen—by considering just a few of these factors, so when you are looking for the levers to change a problem in the world, generally there are just a few really big ones that make most of the difference. We call these *first-order* factors. Once you understand the truly important parts of a problem, you can go back and assess the impact of a few more minor (*second-order*) causal factors, if you want a little more precision in your predictions; but even then you want to use the next-most-important levers and not get distracted by the rest (e.g., *third- and fourth-order* factors).

To make this a little more concrete, consider the simple-sounding problem of navigating from one place on Earth to another. You can understand a lot about navigating on Earth by just treating it as a sphere. It's a pretty good sphere. So the first-order explanation for many things about traveling on Earth—like what we see when we look toward a horizon—is that Earth is a sphere. Now, that order of explanation works for many purposes (much of air travel can be navigated this way), but it's unlikely to give you enough information if you're planning to drive across the United States. In that case it is probably helpful to know some second-order factors about Earth, such as that there are wrinkles on its surface—things like mountains. If you relied on the first-order description of Earth as a smooth sphere, you'd be very surprised when you came to the foot of the Rockies.

Often it takes a little more delving to identify what the first-order causal factors are versus the second-order or less important factors. Imagine that you are trying to figure out the factors that most determine why one person would earn a higher salary than another. Say Joe is a high earner. Is this because he's a hard worker? Our first thought might be that hard work is probably one of the most important, first-order, factors governing his earnings. Another might be his profession. But then somebody might point out that it also seems relevant whether or not Joe lives in New York or the Gobi Desert. It certainly seems easier to be a high earner in New York than in a desert. In fact, there are probably many places in the world where it would be hard to earn as much as Joe. If the variation in earnings *between* locations is much larger than the variation among workers *within* locations, then location would be considered the first-order cause, and hard work and profession second-order causes.

Do you think Canada's location, adjacent to America, is a first-order cause of large amounts of Canada-America trade, a second-order cause, or a third-order cause? At first, you might lean toward not putting Canada's propinquity as a first-order cause; possibly you were thinking that the first-order cause would be that Canada is a highly industrialized country with very sophisticated manufacturing, and that its location is second-order. But then you learn one more fact: Mexico is apparently our third-largest trading partner (right after China). If Mexico is third, despite a business environment that is much less industrialized and sophisticated than Canada's, you might start to think that Canada is indeed getting a big benefit by being so close. So perhaps propinquity is a first-order cause after all.

Understanding the world in terms of first-order, second-order, third-order, etc., causal factors has made it possible to achieve the huge scientific advances and consequent technological leaps that shape our world today.[67] Even the simplest predictions of the motions of objects require us to ignore all sorts of second- and third-order factors, like air resistance and the pressure due to sunlight (this last factor is perhaps only fourth-order generally, but first-order for the motion of certain objects in space). But it is perhaps even more obvious how these orders of understanding play a key role in determining whether any and every decision we make is effective or not. If we get sidetracked by the second- and third-order causes, we can't effect change; we need to identify and focus on the first-order causes.

Sometimes we are good at doing this kind of orders-of-understanding triage quite naturally. When your car's steering suddenly starts vibrating and pulling off-center, you look for a flat tire as the probable first-order cause and don't worry about the second-order cause of aging shock absorbers or slight misalignment of the suspension. But sometimes we chase down all sorts of second-order explanations and miss the big first-order one. We might look for subtle explanations for why our mood or our child's mood is down (Is my job a perfect fit? Is my son's teacher understanding him?) and miss the biggest factor, lack of sleep. Clearly, tech support help desks know that we don't always focus on the first-order source of a problem: they begin the call by asking if our product is plugged in! (Come to think of it, most of us would truly benefit from a good orders-of-understanding analysis of the causal factors of a good night's sleep. We could stop alternating between different remedies for insomnia: exercise in the day, meditation and stretching at night, timing our last meal of the day, and clearing our conscience by doing good so we don't have to answer "How do you *sleep* at night?!" accusations.[68])

On a bigger scale, suppose you're on the town council and you want to reduce the number of local traffic accidents. Obviously, it's crucial to know what all the first-order causes of traffic accidents are. These are things on which you'd want to focus your policies, your new laws and regulations. You'd probably want to reduce drunk driving, distracted driving, speeding, and drowsy driving. And you'd first want to do studies to confirm that these are all first-order causes, or perhaps to discover that one of these is much less of a concern than the others, a second-order cause. You wouldn't bother to regulate driving while blindfolded—this would clearly cause accidents, but it's not one of the first-, second-, third-, or even fourth-order causes of the current accident rate. (An important subtlety: When you're looking at what's the first-order cause and what's the second-order cause, it really depends on how you frame the issue. For example, if you ask someone developing self-driving cars about the first-order causes of traffic accidents, they might say that in ten years we'll take it for granted that the only first-order cause is having humans at the controls.)

Continuing on this theme of your presumed future career as a legislator trying to solve societal problems: If you're in Congress and want to reduce the deficit, and people are saying that spending on entitlements like Social Security is the main cause of budget deficits, you would want to ask—and find out—if that's really a first-order cause of the problem.

Or maybe a fellow senator will say, "Well, you know, we're spending all of our money on those peanut farm subsidies," and you'll respond, "Peanut farm subsidies? They aren't even third-order causes of the deficit!"

TROUBLE IN PARADISE

There's something a little bit magical about a world that allows us to parse problems into orders of understanding. It seems lucky that if you just look for the dominant, most important causal factors for something that you're trying to explain, that usually gets you far. And then if you consider the most obvious exceptions to those simple first-order explanations—these exceptions are your "second-order explanations"—you can usually go home, you're done. The first-order and second-order explanations are all you need, much of the time; you don't have to hold 127 different variables in your head to figure something out. Now, of course there are probably many things in the world that we don't understand, where presumably the reason we don't understand them is that those cases *would* require us to hold 127 variables in our head because they are all comparably important factors. But it's still remarkable how much we are able to figure out about the world using this method.

There *is* a catch (there is always a catch!), which may have left you feeling vaguely dissatisfied with the orders-of-understanding story so far: Nobody is telling us ahead of time which factors are the first-order factors, and it's not always obvious. Not only is it not always obvious, but our first guesses are often wrong.

Let's consider again the topic of government budgets, since as citizens we are often called on to referee (by voting) the arguments between political candidates about what is really important. Imagine that a candidate was arguing that we could dramatically improve the long-term viability of Social Security or the quality of education by reducing incarceration. As a starting point, try writing down—just based on your general knowledge or impression—how you would rank incarceration, education, and Social Security in terms of their contribution to total government social spending. (Assume that we're considering all spending—federal, state, and local.) Which one do we spend the most on, which second most, and which least? In fact, save the piece of paper on which you wrote this. We'll come back to it in a minute.

What is your confidence level in your ordering of the budget costs for incarceration, education, and Social Security? It's not trivial to differentiate the first-order from the second-order factors in a problem like this one, so your confidence may not be very high. Here is where we introduce the next crucial tool from our scientific thinking toolbox, *Fermi estimation*. It's not a panacea—it won't solve such problems every time—but it does get us on the right track and can save us from mistaking second-order for first-order factors, and then following false leads that can waste years of effort or huge expenses. (And don't worry if "Fermi estimation" sounds dauntingly technical—it really isn't, and indeed you may discover that it's something you've already done in the past.)

THE CAN-DO SPIRIT PAYS OFF

The idea of Fermi problems goes back to the famous physicist Enrico Fermi, who was always challenging his students with problems that required quick, on-the-spot estimations. Presumably, Fermi didn't call these Fermi estimates, but ever since, generations of physicists have been challenged to give rapid estimations to answer these "Fermi problems."[69] A famous example from Fermi was: How many piano tuners are there in Chicago? (Let's use this Fermi problem as a challenge for us authors: By the end of this chapter we should leave the readers in a position to answer it themselves.)

Fermi estimates serve the very practical function of helping us differentiate first-order explanations from second-order. And in a world where numbers are thrown around to prove points, it's also a remarkably useful technique for checking whether those numbers make any sense. But Fermi was probably also interested in the can-do spirit that the students learned when they became comfortable with this rapid estimation approach. It ends up feeling quite empowering to realize that we can get a handle on the world in this way.

Before we apply Fermi's estimation approach to our government social spending estimates, let's start with a different, somewhat simpler, example. Imagine that you're trying to estimate how many cars there are in the United States. You don't have access to the internet, so you are limited to knowledge you have in your head. The first step is to ask yourself how you could get at this estimate based on other estimates that might be easier to do because they are more familiar. You try out a few ideas and quickly discard them because they don't seem easier: e.g., estimating the miles of roads in the US and how many cars there are per mile of road, or estimating the number of cities in the US and how many cars there are in each city. But then you think you might try estimating the number of people in the United States and then what proportion own a car, since many of us have *some* idea of the population of the US, and the car ownership patterns among our friends and acquaintances can help us estimate the number of cars per person.

You might remember that the population of the US surpassed 300 million in recent decades. So you can use the 300 million figure or something a little higher to account for population growth since then. Let's use 330 million, 10 percent higher.

If everybody had a car, there would be about 330 million cars. But not everybody has a car. Babies and children don't have cars. Many young adults don't have cars. Lots of older people share cars. On the other hand, some people have more than one car, and some families have almost as many cars as family members. So maybe there are about half as many cars as people. This would suggest that there are 165 million cars out there on the road. We are probably off by millions or tens of millions of cars, but we can be pretty confident that we have the right number of digits in our estimate. That's what we mean by doing this *approximately*. You're not trying to get a final, accurate number. You know that's not possible to arrive at, given the information at hand. (By the way, we looked this up, and we're within 30 percent or so. Perhaps better depending on what vehicles you count.)

When you are unsure of what numbers to use for your estimates, it can be helpful to establish upper and lower bounds. It could be, for example, that you just don't know how many people there are in the United

States, but you know enough to give an upper bound and a lower bound. You can start by testing the plausibility of round numbers. There are certainly more than 100,000 people in the US; are there more than a million? If that sounds good, think about the upper bound. Could there be more than a billion people? No, that's sounding a little much. So you can guess that the population is somewhere between a million and a billion. For many purposes, that's all you need. (You can already confidently reject the clickbait claim that a national "send-a-dollar-bill" campaign in response to a hard-luck story created an instant billionaire.)

Here's a quick example of when you might use both the upper and lower bound. Suppose you wanted to know how much money Americans spent on gasoline for their private vehicles in the previous year. Ask yourself... Do you know that it must be more than $10 million? Do you know that it must be more than $100 million? Do you know that it must be more than $1 billion? What is the highest number that you are sure is still lower than the actual number? This would be your lower bound.

Now let's go in the other direction. What's the upper bound? Is the amount definitely less than $100 trillion? Yes. Is it definitely less than $10 trillion? Yes. Is it definitely less than $1 trillion? Probably. If we establish a lower bound of $1 billion and an upper bound of $1 trillion, that may be as far as we need to go. It would already be apparent that this amount could be a significant part of America's annual imports, so these lower and upper bounds might encourage us to continue with refining our estimate.

TRICKS OF THE TRADE

In these first examples, we have used three useful tricks for making Fermi estimates, which are worth listing:

Estimate in familiar terms. Break down unfamiliar, less accessible quantities into familiar, more accessible ones. (In the case we began with, we had a better feel for the US population than the number of cars, so we based our estimate of the latter on the former.)

Be approximate. Estimates are approximations by definition, but it's important to recognize that it is usually okay to be "close enough." Often you are looking for an answer that is good to within a factor of three; in other words, if the true quantity you are trying to estimate is 100, then any estimate between 33 and 300 is often good enough. This means that the familiar numbers that you are using to build up your estimate don't need to be more accurate than that either.

When you are unsure, *estimate upper and lower bounds first.*

We'll try taking these estimation tricks a little further with the question we just asked: How much *did* Americans spend on gasoline last year? Let's try to get it a little bit more precise than that big broad range of $1 billion to $1 trillion that we just worked out.

Okay. Let's begin by breaking the question down into the components. We already started the ball rolling by estimating how many cars there are in the United States, so let's start there. If we know how many cars there are, maybe we could use that together with how many miles a year the typical car drives. Then we can estimate how many miles per gallon a car gets. And then how much gasoline costs per gallon. These kinds of things you may know already, and you can put them all together to arrive at a reasonable estimate.

So let's go through this list of ingredients as best we can. We've already estimated that there are 165 million cars in the US. Roughly, how many miles does a car drive per year, on average? If you've ever tried to buy a used car, or if you've ever seen anyone selling or buying a used car, you may have a feel for that. Generally, a reasonable used car will have traveled not much more than 10,000 miles per year (perhaps 12,000?). And 10,000 miles is a nice round number, so let's go with that. Now gas mileage. How many miles per gallon does a car get on average? Lots of trucks and SUVs get under 20 miles per gallon, but hybrids get over 40. Let's say the average is about 20 miles per gallon. For our purposes, we need to know how many gallons are used up per mile, so we really need the reciprocal of that, which is 1/20 or .05 (gallons per mile).

So far, our numbers put together will tell us how many gallons all those cars use in a year, so now we need to know how much gasoline costs per gallon. What's a good number? Out here in California, it was typically $4 per gallon when we were writing the book. Let's use that number and just remember that it might be a little higher than the rest of the country. Now let's put the whole kit and caboodle together and see what we get.

165,000,000 cars · 10,000 miles/year/car · 0.05 gallons/mile · $4/gallon
= $330,000,000,000/year

Our estimate is $330 billion. That falls within the range set by our lower bound of $1 billion and our upper bound of $1 trillion.

Having accomplished this Fermi estimate, we can now also say something about our credence level, our confidence in this estimate. Look at all the numbers we used to arrive at our estimate and determine to what extent you would have been willing to go significantly higher or lower. This is a way of estimating how confident you are in the number we reached. Back in chapter 4, we played a game asking you to state a confidence level for any proposition of fact that you make. Now, after working through the Fermi estimate, most readers will probably be starting to feel reasonably confident (say even 80 to 90 percent confident) that the actual amount Americans spend on gasoline per year is between $100 billion and $1 trillion. Of course, you wouldn't want to bet your house on it, which is good. This seems like about the right degree of confidence. (In principle, we could also say how confident we are that the actual number is within a narrower range for which our estimate of $330 billion is the midpoint. For example, we might still be 70 percent confident that the actual amount is between $200 billion and $460 billion.)

In this particular case, the answer is, in fact, between $100 billion and $1 trillion. When we looked this up, we found that Americans spent about $400 billion on gasoline back in 2012 (and about $560 billion in 2022 when the prices were high). So we actually did pretty well in our Fermi estimate. If you estimated that we spent $330 billion a year for gasoline, that might be just as good as knowing that we spent $400 billion. For example, imagine somebody says to you, "With my new invention, I figured out a way to make cars five times as efficient. I'm going to save the United States ten trillion dollars next year!" You would look at them and say, "No, you're not. We don't spend that much money on gas in a year. You can't possibly save the US that much money." (We won't always be that annoying person who pops somebody's bubble—sometimes our Fermi estimate can back them up!)

YEAH, BUT FERMI DIDN'T HAVE THE INTERNET

At this point it's probably worth mentioning that this whole game of Fermi estimations has changed dramatically since Fermi's time because you can now look up lots of numbers online. So many facts and pieces of data are available with a few touches and swipes of your screen that, in practice, your starting point for a Fermi problem might be very different than it would have been years ago.

But in the digital age, there are three reasons why Fermi estimation is still useful. First, you still have to figure out *what to look up* online as a starting point for making an estimate of a quantity that you might not be able to find directly. And that's similar to determining what kind of things you might already know that you could use as the basis for an estimate. Second, as you go through life and come across claims online (and elsewhere), your job is to ask yourself, "Does that make any sense at all?" People are going to claim all sorts of things on a website about how this thing is huge, or that thing is tiny, and your job as a Fermi estimator is to constantly say, "Wait a second. That can't possibly be true because (e.g.) I know there's only this number of people in the world so that number makes no sense." You should be using Fermi estimations all the time to check the numbers you hear, just to make sure they make sense. This is a physicist's "sanity check." Finally, the process of thinking through the Fermi problem also forces you to analyze what the problem is composed of, and that in itself is a key step to understanding what might be first order factors; the process can also help you remember factors that are not first or second order, but which you want to keep in mind.

PUTTING THINGS INTO PERSPECTIVE

Finally, let's return to that orders-of-understanding example concerning which type of government spending is a first-order factor in the national social spending budget and which is second-order among just this subset of expenditures: (1) incarceration, (2) education, and (3) Social Security. This question brings up an extra trick that is helpful: Since you are estimating three different quantities for the purpose of comparing them, it is important to try to do them in a similar way so that you can compare the ratios. That way, even if your estimates are off, the ratios may still be right. Here's a suggested strategy: In each case, estimate what fraction of the US population (estimated above) is the target of either education, incarceration, or Social Security. Then estimate, for each person in that subset of the population, how much it would cost to either educate them, incarcerate them, or socially secure them. This is best done in a table so that you can easily compare those ratios.

Estimates of how many people per year are the target of each kind of government spending are not very difficult. With a life expectancy around eighty-five years, only a quarter of the population or less is old enough to be on Social Security. A roughly similar proportion is in school. The number of people in prison is much smaller—certainly in the single digits in terms of percentage. You may have heard that the United States has one of the highest rates of incarceration in the world; you have a very rough sense for how common incarceration is in your own community; but you also know that it varies strongly by gender, locale, age, socio-economic condition, and race. So after weighing these considerations, let's use, say, 2 percent of the population as our starting guess for the fraction of the population in prison (that's one out of every fifty people, which sure sounds like a lot!).

The costs per person per year may be a little more difficult to estimate. What costs are involved in education? There are teacher salaries, administrator salaries, janitor salaries, building maintenance and utility costs, the capital costs of building schools, the cost of buying supplies and textbooks. As a start, we could assume that an average elementary school teacher makes $50,000 per year and has twenty-five students. That means each student accounts for about $2,000 of the salary cost. Perhaps tripling that number would cover the other salaries and the other costs. Maybe that's too low, but again, the point is to ballpark it.

For incarceration, there are similar kinds of costs, but teachers are exchanged for guards, and schools for prisons. There are no textbooks (at least for many prisoners), but inmates have to be fed three times a day and clothed. And they need medical care. Comparing the two, you have to conclude that it costs much more to keep people locked up twenty-four hours a day than it does to educate them for six or seven hours a day. Let's say the cost is three times as high, or $18,000 per prisoner per year.

Estimating the annual cost of Social Security per person is easier: you may already know, for example, the size of your father's or grandmother's monthly Social Security check. Or you can go by your knowledge that Social Security is intended to provide a retired person with the minimum amount one needs to live on. In most parts of the country, that would be about $20,000 a year. Inserting these rough but plausible estimates into a table, we get the following:

Fermi Estimates for Government Spending on Three Social Functions

	Percentage of population involved during any year	Number of people involved	Cost per person per year	Total annual cost
Education	25%	80 million	$6,000	$480 billion
Incarceration	2%	6 million	$18,000	$108 billion
Social Security	25%	80 million	$20,000	$1,600 billion

You can see from the entries in the table that you don't even need to do the math for the totals in the right-hand column to do comparisons. The estimated annual costs per person for incarceration and Social Security are very close, but we estimated that there are far more people on Social Security than in prison, so more would be spent on Social Security than on incarceration. Similarly, the numbers of people involved in education and in Social Security are estimated to be about the same, but less is estimated to be spent per person on education. Comparing education and incarceration, you can see that the number of people involved in education is perhaps more than ten times higher than the number in prison, while the per-person cost of incarceration is probably only about three times higher than it is for education. That means more would be spent on education.

So already, without calculating the totals, we with some degree of confidence the relative ordering of the three government functions in terms of spending: Social Security costs the most, followed by education and then incarceration. The actual numbers show that we are right. Annual taxpayer-funded spending for one recent year is about $800 billion per year for education, $60 billion for incarceration, and $1,100 billion for Social Security; none of our final estimates are off by more than a factor of two.

Have a look at what you wrote down for your guess at the relative ordering (assuming that you actually went along with this game) and compare it to the ordering from the Fermi problems—which is, apparently, the true ordering. We have found that most people initially guess that more is spent on incarceration than on education, but of course our Fermi problem approach showed that this is very unlikely to be the right ordering. In fact, the Fermi estimate makes it quite unlikely that incarceration could be a first-order factor at all. And, recalling the start of this question from several pages back, it becomes clear that the political candidate's proposal to dramatically change Social Security or education funding simply by reducing incarceration isn't likely to work.

Of course, these cost estimates and their relative ranking in a government budget are only useful as we try to compose our picture of the whole social spending budget (and perhaps tell us if a political candidate is misstating the current situation). They do not tell us what the relative rankings *should* be, just what they are. (Later in the book we will look at how we can also effectively consider what *should* be—which brings in questions of values.)

There is one last important point to make about Fermi approximations: If you're not someone who is doing small calculations like this all the time, the prospect of using a Fermi estimate might not sound that appealing; and yet if you can get over the discomfort, the astonishing thing is how far it can take you. We have seen many people genuinely excited by this experience. If you haven't already done such estimates, you should try a few just to see how satisfying it can be.[70] The goal is to come away feeling so empowered by this that whenever you find yourself feeling down, for whatever reason, you can cheer yourself up with the knowledge that you can go out there and get a handle on the problems of the world with remarkably effective estimation methods!

Scientific optimism, orders of understanding, and Fermi approximations together provide us with something of a can-do tool set with which to take on those big (and small) problems of the world. You should find yourself thinking of them, for example, any time you hear a number bandied about, or a claim that something is a more important factor than anything else. For such a claim of priority, you should immediately find yourself asking, "Okay, what are all the other causal factors and do I have reasons to believe they're out of order?" And if somebody says to you, "We're never going to figure this out. There are way too many factors in this situation," you should find yourself saying, "Well, there could be millions of reasons why, for example, illiteracy rates are going up in this population, but that doesn't mean that we can't find one or two leading first-order causal factors and work on them to make some iterative progress. Let's try some estimates!"

Part IV

Minding the Gaps

CHAPTER 12

Why It's Hard to Learn from Experience

In this fourth part of the book, we turn to look at the some of the particular and surprising ways that individual human thinking goes wrong. Let's anchor this discussion by quickly recapping where we are in the book. In the previous three parts, we have been discussing a broad range of science-sprung thinking tools, with at least three specific goals in mind: First, every one of us could—should!—be using this panoply of tools all the time to make our day-to-day decisions and plans; these tools are generally protecting us from a variety of ways that we otherwise will fool ourselves and are empowering us to take on the complexities of the world. Second, a basic familiarity with these tools helps make sense of (and sometimes test) what we are hearing from scientists, doctors, and other researchers who provide us with key information needed in decision making. Finally, when we try to identify experts who can summarize findings about the world, we can look for their understanding of these tools as a hallmark for more trustworthy sources. If we authors have been doing our job well, you should have been seeing all of these uses in the past chapters.

But it's natural to find ourselves thinking, Why do you need these sometimes elaborate thinking tools? Can't we just learn what we need to know about the world from experience? In the day-to-day world of raising children, collaborating with coworkers, cooking dinner, or voting on local initiatives, the relevant variables are right under your nose (literally or figuratively), detectable without special equipment. With a little trial and error, you can work things out, right? After all, we quickly learned not to touch a hot stove, and we did so without any specialized instruction from experts.

It is hard to shake the impression that we are good at learning from experience. It is common in English-speaking countries to brag that we graduated from "the school of hard knocks." The philosopher John Dewey famously claimed that "all genuine education comes about through experience," and his ideas remain influential through the popular experiential learning movement in education. When we need a surgeon to fix our body or a contractor to fix our house, many of us may prefer the most experienced practitioner.

Nevertheless, there is considerable evidence that experience is often a disappointing teacher. For example, employers place a lot of weight on work experience when choosing among job applicants, and yet work experience is a surprisingly weak predictor of job performance.[71] Yes, experienced workers outperform complete novices, but very experienced workers do not necessarily outperform workers who have gotten past the initial training phase of the job. In many specialized professions like medicine, older workers who are still at their peak mental and physical ability nevertheless fail to keep up with new developments in their field. They stop learning.

When we look at the history of innovations in technology, medicine, and science, it is striking how many discoveries and inventions required no advanced equipment, complex mathematics, or large sums of money. Examples include the lever, phonetic alphabets, the nail, the assembly line, and controlled experiments. So why did they take so long to arrive? For example, the germ theory of diseases was first proposed in the 1500s, but it took another three centuries for Louis Pasteur and John Snow to convince others to take it seriously. Given that our brains had already reached modern form long before recorded history, it seems possible that many such innovations could have occurred far sooner than they did.

And consider just how little of what you do today would be possible if you had to start from scratch. Would you know how to make yourself durable and comfortable shoes? Design a toothbrush, eyeglasses, or duct tape? Discover that brewed coffee beans can make waking up easier? Our kids ask in amused despair, why haven't Mom and Dad discovered even the most basic features that were updated to their phones last year?

One reason is that many features of the environment make it hard to learn from experience. Our senses are bombarded by stimuli, but in the language of chapter 6, some of them are useful signals but others are random noise. Some of those signals are correlated with others, but it is hard to discover reliable relationships in the world because they are often probabilistic ("A will probably be followed by B") rather than deterministic ("If A, then B").

"Okay," you respond, "maybe I won't get it right the first time, but I will figure it out eventually through trial and error, right?" But trial and error is extremely difficult when the environment keeps changing—thus the familiar criticism that someone is too busy "fighting the last war"—or when outcome feedback is often delayed (sometimes by months or years). And because relationships are probabilistic, sometimes good outcomes will follow wrong actions, or bad outcomes will occur despite optimal actions. Making matters worse, our efforts at trial-and-error learning rarely even approximate a controlled experiment. We rarely try to isolate one variable at a time, holding everything else constant. We rarely get a chance to observe the counterfactual—what would have happened if we'd done B (or nothing at all) instead of A.

But many of the factors that make it hard to learn from experience are psychological rather than environmental. In this chapter, we will consider some of the main ones.

At the outset, we want to make clear that none of these factors distinguish "laypeople" from scientists and other experts. We are all prey to these influences, on a daily basis. Scientists are not less vulnerable, and we will provide ample evidence of that in the next two chapters. As we'll explore in a later chapter, what allows science to (sometimes) avoid these psychological factors has less to do with characteristics of *scientists,* and more to do with how *scientific methods* and habits of mind can help us to overcome our limitations.

We should also emphasize that the psychological factors we are about to look at are not "pathologies" but properties of normal human cognition. As such, it is likely that they are pervasive because they are adaptive in some way. Most of them involve ways in which our brains can function efficiently in situations where more systematic reasoning processes are difficult or taxing.

HABITS

If you consider all the skills you've acquired in life, you may notice that most of them can be performed with little or no conscious deliberation. When was the last time you needed to think about how hard to press an accelerator pedal, or turn on hot water, or tie your shoes? These are all *habits*. Habits allow us to multitask, but they also save energy, because it is exhausting to consciously think about every single thing you do. (If you've ever attempted Buddhist mindfulness meditation, you'll know what we mean.) Our brain's ability to "automate" our performance frees us up to turn our attention to new things, like the story a passenger in our car is telling us. Habits are so essential to normal functioning that William James called them "the great flywheel of society." Alfred North Whitehead argued that "civilization advances by extending the number of important operations which we can perform without thinking of them."

Habits aren't entirely unconscious, but they are very hard for us to observe and control because they occur so quickly and so effortlessly. When we first develop a skill, we have a chance to observe its effects on outcomes we care about, allowing us to adjust or reject it. But as a skill becomes more automatic, it becomes more difficult for us to carefully monitor how well it is working for us. "Bad habits" are habits that have become dysfunctional, but because they are so effortless it can be hard to break them. Thus, habits can keep us from learning from experience because they allow us to act without paying attention.

HEURISTICS AND BIASES

Another set of factors that make it hard to learn from experience involves various biases in our judgment—biases that can make us overlook, distort, or deny important information in our environment. Like habits, biases can often result from the way our brain tries to do things quickly, without a lot of attention. The cost comes when we fail to make the best use of the evidence at our disposal.

The word "biased" probably gets thrown around too casually. It is easy to accuse someone else of being "biased" just because we don't like that person's point of view. Fortunately, as we discussed in chapter 9, judgmental biases can be defined fairly objectively. Recall that a process is *noisy* when it makes lots of random errors, but we judge that a process is *biased* when it makes systematic errors—either consistently above or consistently below the correct answer. So when we have an objective criterion or true value, we can identify someone's biases by comparing their responses to that criterion or value. When we don't know what is objectively true, that approach won't work, but researchers have developed a variety of experimental strategies for rooting out biases.[72]

Wikipedia offers a list of documented cognitive biases,[73] and the last time we checked, the list had 123 entries! This might give the impression that psychologists get paid by the bias, and perhaps some do. Many of these biases were first documented by the psychologists Daniel Kahneman and Amos Tversky. Tversky passed away in 1996 at a young age, but Kahneman later received a Nobel Prize for their joint work. Kahneman's bestselling book, *Thinking, Fast and Slow,* is the best introduction to these biases, and if you haven't read it, we urge you to do so.

Some of these phenomena have the word "bias" in their name (e.g., confirmation bias). Others are called "heuristics" (e.g., the availability heuristic). The distinction between the two words is fuzzy, but roughly, the label "bias" describes the outcome (a systematic departure from the true value to be judged), whereas "heuristic" refers to the process that generates a particular bias. A heuristic is a specific rough-and-ready (but fallible) method we've learned for making quick judgments, in order to avoid the hard cognitive work of careful deliberation.

The biases can vary along a continuum from "hot" to "cold" (to use a temperature metaphor). Hot biases are easiest to describe because they are so familiar to us. They emerge because of emotions (especially anger or fear) and motivations (what we want to happen or what we want to believe). At the other extreme, cold biases seem to be by-products of our typical ways of making quick judgments, even in situations where we are calm and cool and collected with no particular goals or desires. An example, demonstrated by Gerd Gigerenzer and his colleagues, is the way we infer the relative size of cities from our familiarity with their names.[74] This usually works because big cities are more famous than small cities. But it can lead us astray; e.g., most people will guess that San Francisco is larger than San Jose, even though their populations are about 815,000 and 983,000, respectively. Why? "Do You Know the Way to San Jose" is a catchy tune, but there are lots more cultural references to San Francisco in popular culture. (It helps to have steep hills, streetcars, and the Golden Gate Bridge.)

We won't spend a lot of time describing hot biases, not because they are unimportant, but because you are undoubtedly quite experienced at spotting them. We've all seen people who seem blinded by their emotions and desires. But though it's hard to admit, we've all *been* those people at one time or another. The reality is that typically, cognition is *motivated* in the sense that there are some beliefs and outcomes we want to be true, and others we hope are not true. Hot biases are often more harmful than cold biases, because people can go to great lengths to distort evidence in order to get what they want, and they may disparage others who reach the opposite judgments. This can lead to intractable conflicts when people insist that they are being neutral while the other side is biased. We have talked to many colleagues who give expert testimony at trial. Most of them will agree in the abstract that many experts are biased, and that being paid by one side creates a potential conflict of interest. But most of them also insist that they aren't the kind of people who would be swayed by money.

We won't attempt to review the dozens of biases that have been documented in the psychology literature. Instead, we'll focus on the ones that seem especially likely to keep us from learning from our experiences.

The Availability Heuristic

This one is mostly a cold bias. As first defined by Kahneman and Tversky, the availability heuristic refers to our tendency to "assess the frequency of a class or the probability of an event by the ease with which instances or occurrences can be brought to mind." Kahneman and Tversky talked about various ways some things grab our attention: e.g., concepts that are linked to lots of other familiar concepts in our memory, events that happened more recently, experiences that are particularly vivid, or situations that are easy to imagine. Here's a simple example from their early demonstrations: Is the letter "K" more likely to be the first letter of a word, or the third letter of the word? Last time we asked a group of students, 61 percent said that any randomly chosen word would be more likely to start with a "K," whereas only 39 percent thought it more likely for "K" to be the third letter. The 61 percent majority was wrong. The letter "K" is more often the third letter in an English word than the first. But it is harder to search our auditory or visual memories for words by their third letter, because when we learn a word we hear the "K" sound first, or we read the letter "K" first. Most of us read ABC books with our parents, but they didn't say, "Big 'I,' little 'I,' what has 'I' as a third letter? 'Rhinoceros'!"[75]

Media exposure is a big reason why some things are more cognitively available. A study in the 1970s found that people believed tornadoes were a more frequent cause of death than asthma, even though death by asthma was 20 times more common.[76] But most people would rather read a news article or watch a movie about tornadoes because they are more dramatic (so we encounter more tornado-caused mortality stories).

Here is a concrete example of how the availability heuristic can play into policy debates. Many people believe that juries in personal injury cases give out awards that are obscenely large and capricious. One of Rob's studies looked at national news reporting of jury trials and compared that to the actual court statistics from various studies.[77] In a sample of cases reported in the media, people who file lawsuits won 85 percent of the time. In actuality, the true rate was in the 30 to 55 percent range, depending on the type of litigation. If you win your lawsuit, what's the average size of the award? The median award in the cases reported in the newspaper and magazine articles was almost $2 million. The actual median award for all trials was between $50,000 and $300,000. So the media—and the availability heuristic—can lead you to an exaggerated expectation of what you'll get if you file a lawsuit.

The Anchoring and Adjustment Heuristic

This is another cold bias. When we have to make quantitative estimates, we often don't know where to start. In a famous comedy scene, master British spy Austin Powers's nemesis, Dr. Evil, was awakened from decades of cryogenic stasis and immediately set out to extort the world's governments for "One . . . million . . . dollars!" — not realizing that in the present day that amount is far too modest to be worth the crime.

But Dr. Evil is not alone. We all have difficulty figuring out how to guess quantities. Kahneman and Tversky suggested that we usually approach this task by looking for some arbitrary but salient number to use as a rough starting point; we then make a judgment about whether to adjust upward or downward. But here's the catch: we usually don't adjust enough. As a result, our carefully considered estimate tends to end up too close to our arbitrary starting point. The astute reader will notice that this anchoring and adjustment process is always a risk when we are making Fermi estimates.

Here's another legal example. Rob and his colleagues were studying how people decide on a fair amount for monthly child support payments after a divorce.[78] Because this is a difficult task, people's estimates are all over the map. One way to reduce this variability is to give people a specific value as an example. When people were given $800 as an example, they recommended an award of around $1,000 on average. But when they were given $1,400 as an example, they awarded about $1,300 on average. There is no objective standard for the right amount of child support, but the point is that providing reference numbers is going to be subject to gaming and manipulation by lawyers who want to influence the judge.

Hindsight Bias

There's an old saying, "Hindsight is 20/20." The ophthalmological notation 20/20 indicates that what you can see at 20 feet is what a normal person would see at 20 feet. (Apparently in metric-system countries the equivalent saying is in meters: "6/6 vision.") So 20/40 means you are nearsighted. The idea reflected in the saying is that after an event occurs, it is easy to get the "prediction" right; hindsight is way easier than foresight. The psychologist Baruch Fischhoff proposed a general feature of human judgment: after we learn an outcome, it will seem like that outcome was more obvious than it actually was beforehand.[79]

Here's an example: In the early 1970s, Fischhoff was studying probability judgments, and he chose future outcomes that either seemed very unlikely to him, or very likely. This was during the Nixon administration, and Nixon was a vehement anticommunist, so for an unlikely event, Fischhoff asked people the likelihood that Nixon would go to China on a diplomatic visit before he left office. As it happens, Nixon actually did go to China on a diplomatic visit in 1972, which surprised even foreign policy experts. Fischhoff had the brilliant insight to go back to people afterward and ask them to try to remember what probability they had given for the Nixon-goes-to-China event. He found that people misremembered their probabilities as being higher than they really were. Basically, they thought they'd expected the event to occur—that they "knew it all along."

Hindsight bias creates problems for students who take classes in the social sciences. If I teach you about a new finding in psychology or sociology, you are probably clever enough to come up with a plausible reason why it might occur. As a result, you may conclude that the finding is pretty obvious, and you may imagine you knew it all along. A common way of demonstrating this in the classroom is to give students a short description of a study of romantic relationships. Half the students are told that the study showed that "birds of a feather flock together"; the other half are told that the study showed that "opposites attract." Both of these sayings are "common sense"—but they completely contradict each other. Unfortunately, after reading about the alleged findings, each group of students mostly rates the finding they read about as "obvious," perhaps leaving them wondering why the professor is teaching them such trivial stuff.

There are now many hundreds of demonstrations of hindsight bias. Some are harmless, but others are more pernicious. Years ago, the ex–football player and actor O. J. Simpson was on trial for the murder of his ex-wife. Prior to the verdict, most people, including experts (practicing lawyers and professional gamblers), thought he would most likely be found guilty. After he was eventually found "not guilty," we'd have liked to think that the experts would have said, "Oops, we were wrong." But instead, many of them appeared on news shows "explaining" that Black jurors (the majority of the jurors in this case) tend to be lenient toward Black defendants. The data available at the time showed that this accusation was factually incorrect, and anyway if they really believed this, why weren't they predicting an acquittal all along? So these experts ended up fabricating a myth about the prejudices of Black jurors.

Ingroup Bias and "Badging"

In the 1970s, the psychologist Henri Tajfel developed the "minimal groups" paradigm, in which people are classified into groups using patently arbitrary criteria (e.g., whether they over- or underestimated the number of dots flashed on a screen). Tajfel showed that even such minimal groupings were sufficient to elicit more favorable treatment of ingroup members in the allocation of rewards for an unrelated task.[80]

Social psychologists have found that people's expressed attitudes are sometimes motivated less by their actual beliefs about the truth of a proposition (e.g., a claim about the effect of the death penalty or gun control on homicides) than by the desire to publicly express their values ("I am conservative/liberal"). We think the label "badging" is an apt way of summarizing this phenomenon, and we will refer to it again elsewhere in the book.[81]

Dispositional Bias

When driving a car, Rob often finds himself silently cursing other drivers who drive too slowly while searching for addresses on the side of the road. These people are clearly so self-absorbed they don't care about other drivers. Yet he often finds himself slowing down near a shopping mall to see if it has a store he is looking for. When someone behind him honks, he is bewildered: Can't they see that the fault lies with the idiot developers who designed the mall so badly? It turns out Rob is not unique in this quirky inconsistency. A 1991 survey of accidental injury victims found that in multiple-vehicle accidents 91 percent of drivers blamed someone else, usually the other driver.[82]

Under the rubric of "attribution theory," psychologists have long studied how people explain the causes of their own behavior and the behavior of others. In 1958, Fritz Heider argued that where possible, people tend to prefer to attribute other people's behavior to "internal" causes (e.g., personal traits like "greed" or "intelligence"), rather than "external" causes (e.g., environmental hazards or low visibility). Conversely, when they themselves did things wrong they would tend to attribute the behavior to external causes. By 1977, Lee Ross argued that evidence for this bias was so pervasive that it should be considered "the fundamental attribution error."[83] In later work, he showed that this tendency often amplified interpersonal conflicts, as each party denounced the motives and shortcomings of the other while discounting the powerful situational forces that were contributing to the situation.

In the 1990s, a number of cross-cultural studies began reporting that the bias wasn't so fundamental after all; e.g., people in Asian countries showed a greater preference than Westerners for external (situational) explanations.[84] For that reason, many investigators prefer to label it a "dispositional bias," one that is now seen to be particularly characteristic of Western cultures. Nevertheless, the tendency to "take things too personally" is recognized in Asia as well. An observation attributed to the ancient Chinese philosopher Zhuangzi states that "if a man is crossing a river, and an empty boat collides with his own skiff, even though he is a bad-tempered man he will not become very angry. But if he sees a man in the boat, he will shout at him to steer clear. If the shout is not heard, he will shout again, and yet again, and begin cursing."[85]

Confirmation Bias

We have been saving the best for last—"best" only in the sense that it is perhaps the most important bias we hope to help people overcome. Confirmation bias refers to the tendency to mostly search for evidence that is consistent with a hypothesis, to the neglect of evidence that might be inconsistent. It also occurs when we have all our evidence but give more weight to the facts that support our hypothesis than to those that don't. Along the same lines, when people do encounter unfavorable evidence, they often subject it to greater critical scrutiny than favorable evidence. Confirmation biases occur along the whole spectrum from hot to cold. The hot version is familiar: someone who selectively cites only those facts that will let them win an argument or get their way, or someone who simply denies inconvenient facts. But the cold version is equally important: People often search for and cite facts that support the hypothesis because it just feels like the logical place to start. After all, if we can't find any confirmatory examples at all, our hypothesis is pretty much dead in the water. The problem occurs when we find the confirmatory cases but just stop there. We will come back to confirmation bias in later chapters.

CONSIDERING THE OPPOSITE: A USEFUL TRICK FOR DEBIASING OURSELVES

> I don't think, sir, you have a right to command me, merely because you are older than I, or because you have seen more of the world than I have; your claim to superiority depends on the use you have made of your time and experience.
>
> — Charlotte Brontë, *Jane Eyre*

> Practice does not make perfect. Only perfect practice makes perfect.
>
> — Vince Lombardi

As we've seen, there are a lot of reasons why we fail to learn as much as we might from life experiences. But to be clear, we are not arguing that it is impossible to learn from experience. Science would be a waste of time if that were true. Rather, learning from experience is difficult, and requires us to study our experiences in a careful way. K. Anders Ericsson, perhaps the foremost expert on the development of expertise, has argued, "The effects of mere experience differ greatly from those of deliberate practice, where individuals concentrate on actively trying to go beyond their current abilities.... [D]eliberate practice requires concentration that can be maintained only for limited periods of time."

By teaching you (or at least reminding you) about these biases, have we made you less biased? We wish it were that simple. Studies show that learning about these biases can reduce their influence, but only slightly. Another approach is to pay people for making the most accurate judgments possible—the idea being that maybe people will skip the shortcuts and engage in more systematic reasoning if there's money on the line. But surprisingly, this doesn't always seem to help much either. Even when the stakes are very high, people often fail to overcome even basic judgmental biases.

Probably the most successful debiasing strategy identified to date is called *consider the opposite*—or, in more complex cases, *consider the alternative.*[86] When you have a strong expectation about some future outcome, stop and consider all the reasons why the exact opposite could happen. When you go through this exercise, what you usually find is, yes, you have good reasons for the choice you made, but you also would have had good reasons for the other choice. In chapter 4, we described a discussion activity (using the topic of increased standardized testing in schools) in which participants are asked to assign a level of confidence (e.g., "75 percent") to every statement they make that could be true or could be false. One of the interesting outcomes was that this led naturally to participants "considering the opposite" as they were made aware of their less-than-99-percent confidence in most of their own statements. "Consider the opposite" isn't routinely taught in science classes, but the reality is that it is baked right into most science methodologies. For example, random-assignment experiments are designed to probe what would happen in opposite ("counterfactual") circumstances.

In the next chapter, we will see how motivated cognition can distort scientific practice, even when investigators purport to be honest brokers of the facts. Then we'll look at some tricks of the trade scientists have developed to try to overcome their biases.

CHAPTER 13

Science Gone Wrong

In 1988 an established director of a French laboratory and his research team published a paper in the highly respected science journal *Nature* that made an extraordinary claim. Apparently, they found that after multiply diluting a solution that originally contained a particular antibody (in fact, diluting it with water as much as 10^{120} times!), the now-quite-pure water that resulted from this over-the-top dilution activity still showed evidence of having some of the same reactivity as the original antibody solution. This, despite the astronomically small odds that any of the original solution could still be present in the water. The paper's suggestion was that the molecular organization of the water was somehow keeping a memory of what had once been present in the sequence of past dilution events.

What are we to make of such a journal article? Well, clearly, since this chapter is titled "Science Gone Wrong," you can guess that the end result here is not going to be an enthusiastic endorsement of a great discovery. But this story gets at a big problem that we all face. There is no label on a science journal article or related news article that tells us which articles cover exciting new results and which ones exemplify science gone wrong. And it can really matter. If we have a loved one who is suffering from a difficult disease, an article suggesting that water can hold a molecular memory might offer us some hope for proffered homeopathic cures, since homeopathy claims results based on similarly astronomically micro dilutions. And if this article is misleading us (we will get to why we think that is the case here), it may lead millions of people to waste money, and worse, threaten their health by neglecting genuinely effective treatment options.

The problem of recognizing good and bad science is even more complex because we're really talking about a whole realm of different kinds of circumstances in which science does not live up to expectations, or in which people are clothing something clearly bogus in the mantle of science in an attempt to give it legitimacy. These circumstances can range from the unintentional and honest to the deluded and fraudulent.

THE GAMUT OF WAYS SCIENCE GOES WRONG

As a reference point, let's begin with good science. Ideally, the science is done well and gets correct results. This is what you really hope to find when you read a newspaper article about a scientific result or when you read a scientific paper. Some good science, however, will happen to get the wrong result. In fact, if you think about our chapter 4 discussion of confidence levels, some good science *should* get the wrong results. A good scientist should give you a confidence level, that is, the probability that the result is going to be right. That's all you expect of them. But if they are giving you a 95 percent confidence level for that result, and if they are doing their best job, they should be wrong in one in every twenty papers. That means there should be good science out there that reports the wrong result; at least one-twentieth of the papers should be in that category if they've been using 95 percent confidence levels.

Then we get to bad science. The most innocent kind of bad science is done by scientists who are doing most things right but do something important wrong. For example, there are papers out there where the authors didn't understand the look elsewhere effect that we discussed in chapter 7: when a study examines more variables than it was originally designed to test for, there's a high likelihood that chance connections will be misinterpreted as meaningful results. You all understand that effect now, so you will never make that mistake, but some scientists do not realize they are falling into the trap of a look elsewhere effect, so their papers claim reliability for a result that isn't justified. That is just one example of science not done well.

We should pause here to say a sympathetic word about mistakes in science. A well-constructed scientific study and analysis can be extraordinarily difficult to get right. There are so many mistakes one can make. Throughout this book, we describe some of the big gotchas that scientists have learned to watch out for over the years, but that doesn't mean that this comes naturally to them. (We admit it: you readers of this book—and we authors—*will* still sometimes make the look-elsewhere-effect error ourselves.) One of the whole points of a scientific community that reviews one another's papers and replicates one another's results is that every scientist depends on others to catch the mistakes they will inevitably make.

So we should try not to feel too smug when a science paper is caught out in a mistake—that's how science works. (And those scientists who are particularly good at designing studies in ways that make it easy to catch the errors are especially prized. Of course, those who constantly report results based on errors, or refuse to fix errors when discovered, are less respected.) Moreover, when we read a science paper that has just been published, our first assumption should be that there may still be mistakes in it that will come to light after other studies have grappled with its results—and this is even after the paper has undergone the often-rigorous scientific review process that each journal organizes.

Next on the spectrum is pathological science. This is a concept that was first presented in 1953 in a lecture by Irving Langmuir, a Nobel Prize–winning chemist.[87] Langmuir described a number of examples where a scientist begins doing real science, but eventually falls in love with a surprising result and starts to ignore all indications that it's a mistake. Here, the scientist goes off the deep end: It's not that they are missing a subtle mistake, as in the previous cases where scientists try hard to figure out any way their result might be wrong. Rather, they are now trying to defend their result against all evidence that it is wrong.[88] This mode of failure by honest scientists is so disturbing that we will return to it after we finish this survey of ways science goes wrong.

After pathological science comes what we sometimes call pseudoscience. People doing pseudoscience use the language of science. They like the terminology, but they apparently don't like the actual activity. So, for example, they aren't really testing to see whether there's a causal connection between what they would like to be the cause and what they claim to be the effect. And they aren't trying to fit their claim into the larger context of everything else we know—what we previously described as the "raft" of interwoven scientific knowledge. They do use a lot of great-sounding words in the process.

Sometimes, it's easy to guess that you're seeing pseudoscience. But often you have to read a little bit more carefully. Maybe the writer formatted the Web page nicely and used nice-looking equations, but as you read it you start to think, Wait a second. The page is using terms that sound like science but the terms are being used incorrectly, and then conclusions are drawn based on this misunderstanding. Moreover, the pseudoscientific page doesn't seriously look for ways that its results could be wrong or show where the current weaknesses are in its own arguments.

Pseudoscience merges into something that's sometimes called cargo cult science. That term for this level of science gone wrong has a story behind it that comes from Richard Feynman, the famous physics Nobel laureate, who introduced it during a talk he gave at the commencement at Caltech in 1974:

> In the South Seas there is a Cargo Cult of people. During the War they saw the airplanes land with lots of good materials, and they want the same thing to happen now. So they've arranged to make things like runways, to put fires along the sides of the runways, to make a wooden hut for a man to sit in, with two wooden pieces on his head like headphones and bars of bamboo sticking out like antennas—he's the controller—and they wait for the airplanes to land. They're doing everything right. The form is perfect. It looks exactly the way it looked before. But it doesn't work. No airplanes land. So I call these things Cargo Cult Science, because they follow all the apparent precepts and forms of scientific investigation, but they're missing something essential, because the planes don't land.

Feynman was using the term "cargo cult science" to talk about science that is a little bit like pseudoscience but possibly worse, because it is only vaguely caricaturing real science, while really doing the equivalent of wearing wooden headphones with bamboo antennae and hoping that will make the airplanes land. Of course, the science doesn't land when you use cargo cult science.

Now, it can get worse than that. At the bottom of the list—the far end of the spectrum—is fraudulent science. Fraudulent science involves an intentional, active effort to mislead others by misrepresenting one's findings. Misconduct can be motivated by the prospect of financial gain, by the fear of losing one's job, or by personal ambition. It is difficult to measure the prevalence of scientific fraud, but while it is far from typical, it may not be all that rare. Surveys of scientists show that almost 2 percent admit (anonymously) to having engaged in fraudulent scientific practices, and about one in seven believe their colleagues engage in fraud.[89] This is scary to see.

Fraud is always bad, but science can still function with occasional fraud because the ordinary processes of science—scrutinizing findings, attempting to replicate them, testing new hypotheses implied by the findings—will probably weed out the fake results, hopefully before too much damage is done.

If somebody engaging in fraudulent science chooses an obscure-enough topic, then maybe they will never get caught, but of course you don't gain much by doing something that's so obscure that nobody cares about it. So we expect—or at least hope—that fraudulent science is a little bit self-limiting. And to the extent that science has been effective as a way to track reality and build efficacious solutions based on that understanding of reality, we can guess that there isn't too much of this out there in the scientific literature. But every few months, another dramatic example of scientific fraud comes to light, amid the million papers published annually.

The interesting thing is that you, the readers of this book, are now training to catch such fraud in a way that is only recently being recognized. Some researchers go back and look at the spread of the distributions of the data that's being reported in papers, especially when they are a little suspicious of a certain result. And they have started to find that when people make up fake data, they're not very good at making fake random noise in the data. When you plot the data, it looks like it came from some perfectly bell-shaped distribution of statistical uncertainty without any of the indicators of the other kinds of noise (including chapter 9's systematic uncertainties) that you might have expected to see for that particular measurement. Or, worse than that, there may not be a bell-shaped distribution of noise at all, in which case all the data looks way too good. It won't even reflect the equivalent of those runs of heads or tails that we know (from chapter 7) real data would have.[90]

What are we to make of this litany of failure modes for scientific claims? Throughout this book we are, of course, highlighting all sorts of failure modes in our thinking, with several goals in mind: We want to recognize such mistakes when we read about science results, and we especially want to find experts who help us catch these mistakes; we want to avoid falling into these mental traps in our daily life (whether or not we are dealing with a scientific question); and we want to avoid being those scientists who make these mistakes in their work. And although most of us are not scientists, it is still very useful—and clarifying—to think of these failure modes from the point of view of the scientists in these exemplar stories of science gone wrong, both those falling into the traps and those working hard to avoid them. (It would be seared into our minds if Mr. Spock went wrong in one of these ways.)

WHAT MAKES PATHOLOGICAL SCIENCE PARTICULARLY WORRYING

Looking at the whole range of problems, it's the middle ones that we probably want to focus on, not the extremes. We can't do anything about good science results that are wrong by random chance (i.e., one in twenty times for 95 percent confidence) other than enlist a community of others to help us find our honest mistakes by, e.g., replicating our results. At the other extreme, most of us are not going to run around using random science vocabulary without knowing what it means—or out-and-out fraudulently making up data to prove a point. But the middle example of pathological science—becoming invested in, or falling in love with, a particular exciting result and ignoring all data that suggests we're mistaken—has many features that we can all recognize as traps we can fall (and have fallen) into.

The Langmuir lecture on pathological science that we mentioned suggested a possible list (which we have slightly reworded) of "tells" that Langmuir saw as giveaways that a science result was likely to be in this dubious category:

1. The effect is produced by a barely detectable cause, and the magnitude of the effect is substantially independent of the intensity of the cause.
2. The effect itself is barely detectable, or has very low statistical significance.
3. There are claims of great accuracy.
4. The work involves fantastic theories contrary to experience.
5. Criticisms of the work are met by ad hoc excuses.
6. The ratio of supporters to critics rises to near 50 percent early on, then drops back to near zero.

So, faced with a worrisome scientific result, one first asks, "Is this effect—or its cause—barely detectable?" Now, of course, if you're worried about pathological science, it's likely that either cause or effect is barely detectable, because if they were obvious, you probably wouldn't be worrying. The next question might be a little more subtle: "Does the magnitude of the effect depend on the intensity of the cause?" This is what we called the dose-response relationship in chapter 3 when we discussed Hill's criteria: you expect that if you take a little more of the "dose," the cause, you should see a little more of the response. That doesn't always happen. For example, a small dose of an antibiotic may have no effect, but when you cross a certain threshold of dosage you quickly get all the effect you ever will, eliminating the infection.

However, when you're already doing something right at the edge of your ability to detect the cause and effect, you need extra evidence that you aren't fooling yourself into seeing a causal relationship that isn't there; and a reproducible correlation between the cause and the effect, a dose-response relationship, becomes an important indicator. This is, of course, a more demanding criterion than just detecting the cause and the effect at all, and it serves as a reminder that, with barely detectable elements, it is really easy to read into your results what you are hoping to see, so you want to be pretty tough on the claim.

Many of Langmuir's criteria appear to be the consequence of people falling in love with their results, to use the language that we introduced pathological science with. In fact, the phrase "falling in love with your results" is actually a very good analogy. When you fall in love with someone, there's that period when anything they do seems so cute and endearing. (You know: "Oh, isn't it wonderful the way they make fun of all my friends?") Then, when all those love chemicals have dissipated and you come back to reality several years later, you wonder, "What was I thinking?"

When you're in love with someone, everything can look good, and this also seems to be the case when you're in love with an idea or a scientific result. You think you've discovered this amazing aspect of how the world works. The world is going to be different because of your discovery. You're excited. You start to throw out data that doesn't give you the accurate result that you wanted, and you only pay attention to the "accurate result" (resulting in Langmuir's point 3). You start developing these amazing, fantastic theories that explain how this must be working (Langmuir's point 4), and you don't think about whether those theories jibe with the many other, more well-established scientific concepts and evidence in the metaphorical raft of ideas we've been weaving over several millennia. It's a coherent story that we build up in science, and when your particular explanation doesn't fit with that story, that should be a red flag; but if you're madly in love with the idea you don't notice the warning signs at that moment.

To make things more difficult, when in love you're inclined to defend your sweetheart against any criticisms from the rest of the scientific community. When others start pointing out what's wrong with your experiment or its results, you start coming up with great excuses. "Oh, yeah, that was just because you came the wrong day. That day, the humidity was high, so that's why the experiment didn't work. If the humidity had been low, it would have worked!" You start inventing such ad hoc arguments to answer any particular criticism (Langmuir's point 5).

There is another clue to be found in the response of the scientific community (Langmuir's point 6). Other scientists are at first excited by the result because it looks like it could change everything in some way. Then, over time, they start getting disillusioned as they try but fail to replicate the result. By the end, basically nobody is buying it.

WHO'S AFRAID OF THE BIG BAD PATHOLOGICAL SCIENCE?

It's probably worth pointing out that this portrait of pathological science that Langmuir sketched out is not a strict or exclusive definition. There isn't something that *is* pathological science and something else that *isn't*. Langmuir's list is just a collection of warning signs: if you are reading something and you start seeing things that look a little bit like this, you should be on your guard. And whether you're a scientist in the middle of doing an experiment or anybody at all trying to figure out a causal connection in order to make a decision, if you start finding yourself doing things that feel anything like Langmuir's list, you should take a day off and ask yourself, "Where am I going with this?" (The goal is not to be the last person at work who realizes that your great reorganization plan just isn't working.)

Because this is a list of warning signs, having just one of the signs be true isn't enough to make a result suspect. Suppose a result has a strong measurable cause and then a strong relationship to a strong measurable effect. Everybody can replicate the experiment and get the same result; people just hadn't made those measurements before. You made the measurements and everybody said, "Wow! You're right! You put more of this in and you get more of that out." The statistical significance was great, and the accuracy didn't need to be amazingly perfect to see it. In that case, even if the result goes against all the other current theories, you take it more seriously. You don't just dismiss it as pathological science. In some sense, that's what happened when both Saul's team and a competing team first made measurements that indicated the acceleration of the expansion of the universe. People were able to see it with a big-enough effect (measured by the brightness of distant exploding stars) that they were able to start to take it in and say, "You know, we're going to have to rethink our combination of interlocking stories that make up our current scientific picture."

When you get a result that challenges current theories, however, the science behind it is scrutinized much more rigorously. It requires a higher standard of proof to accept it. Using our raft metaphor, the challenging result is like a new log that doesn't fit into the raft in any way. You don't want to ignore the log just because you don't have a way to fit it into the whole story, so you hold it in a bit of a quarantine for a while. Eventually, if you have enough logs to put a new and improved raft together around it, you do that. That's what happened when Einstein proposed his theory of relativity, which allows us to consider the idea that space itself could be curved. If there is a bizarre idea like that, one that just doesn't make sense from an ordinary point of view, you don't want to lose it just because it is hard for us to visualize how it works.

WHEN THE RUBBER MEETS THE ROAD

Now that we have the full panoply of science gone wrong laid out in front of us, and now that we stand properly warned about falling into pathological science, what are we to make of the science news that we read, or the latest evidence relied on by a plausible expert? Even for scientists it can be extremely difficult to read, understand, and properly evaluate scientific articles even in neighboring subfields, let alone areas where one doesn't know the vocabulary and the field-specific experimental challenges. But although much of the scientific argument is hard to follow, it is sometimes possible to catch the tone of the work and especially to look for evidence that the authors were sufficiently on the hunt for their own errors and ways that they could be misled.[91] (It's not coincidental that this is very similar to our best prescription, the consider-the-opposite strategy, for mitigating our cognitive biases.) And the Langmuir list often provides us with specific indicators of an article's tone.

SOME CASE STUDIES

Cold Fusion

A dramatic example of science news that calls out for such evaluation can be found in the ever-popular and admittedly important topic of generating energy from nuclear fusion. Every few years there is a flurry of science news articles in which a breakthrough is announced in the quest for fusion as an energy source. It is no surprise that these science claims make the news; if we succeeded in generating large amounts of usable energy from an abundant fuel (e.g., one found in seawater) and created only small amounts of manageable waste product and no additional greenhouse gases, we could dramatically improve the lives of everybody in the world. This ultimate goal has led to two major long-term international efforts using two different technological approaches (each with iterative advances) to learn, eventually, how to create a controllable continuous fusion reactor, and then make this an industrially viable power source. However, along with the slow-but-steady series of news results from these two multibillion-dollar efforts, news articles also intermittently appear from teams that claim to have made a breakthrough that will void the need for these huge, many-decades-long developments. Let's look at one such announcement, the 1989 "cold fusion" claim.

Cold fusion is so called to contrast it with the extremely high temperatures achieved (an order of magnitude hotter than the sun) with football-stadium-size, power-hungry machines that the two major technological approaches require to create fusion. In the spring of 1989, two well-credentialed chemists, Stanley Pons and Martin Fleischmann, gave a press conference in which they claimed to have produced energy from fusion produced in a tabletop-size experiment in which they ran an electric current through a standard chemistry setup (called an electrolytic cell), but specifically with the use of palladium and "heavy" (deuterium-rich) water. The scientific world—and everyone else—was excited at the idea of this experiment by established scientists (they weren't presenting pseudoscience; they understood the words they were using), and immediately started to try to understand how it worked and to check and reproduce the results.

The many attempts to replicate the experiment were generally failing, and other scientists started finding flaws and sources of error in the original experiment. For example, physicists started to point out that if you had the sort of fusion that Pons and Fleischmann claimed was occurring, it would release a lot of strong radiation—enough radiation effectively to kill you if you were in the room—and yet nobody seemed to be hurt. It was also odd that the extra energy release could occur well after the experiment began, without a change in the electric current input into the electrolytic cell. Pons did not respond to the vast majority of the queries, including questions regarding the experimental setup. Fleischmann and Pons remained convinced of their findings, but by the end of the year most scientists considered the cold fusion claims to be dead.

Many of these concerns were covered by the news media. In the *New York Times,* Malcolm Browne wrote: "Dr. Pons and Dr. Fleischmann... refused to provide details needed for follow-up experiments." Robert L. Park, in a book called *Voodoo Science* that looked back some ten years later, said, "At the power levels claimed by Pons and Fleischmann, their test cell would be expected to emit lethal doses of nuclear radiation.... [I]t should have been the hottest source of radiation west of Chernobyl." A *Scientific American* retrospective at this time said: "There is one point on which all true believers in cold fusion agree; their results are not reproducible. To most scientists, this implies that cold fusion results are not believable, but true believers suggest that this unpredictability makes them more interesting!"

Which of Langmuir's criteria are in play here? Certainly, it doesn't appear that the magnitude of the effect (the amount of fusion energy produced) is dependent on the intensity of the cause—the energy from the input electric current wasn't changed when the claimed effect began. All of our physics experience, and the corresponding well-established physics theories, would suggest that fusion should generate testable by-products, like radiation, so the absence of such by-products would at least require some theory that would not fit with the best current "raft of science," if not a fantastic theory. Similarly, the lack of confirming experiments reproducing the result seems to have led to ad hoc excuses (though it is not clear if these came primarily from the original team or from their supporters). And support for the claim rose and then declined as more information came to light. So four of Langmuir's criteria seem to apply. Even though the other two criteria (a barely detectable effect and great accuracy) don't appear in this story, there is enough reason to be concerned about the validity of this experiment that by the end of 1989 you probably wouldn't have wanted to invest your money in a cold fusion company.

It's important to be clear here that the idea of searching for a new and unusual route to fusion that would be much easier to use is a great example of science in action. Even finding a result that turns out not to be reproducible—or to have experimental flaws—isn't necessarily bad. But what we all aspire to is the ability to step back and look proactively for the errors. Doubling down is not a virtue here. It is the apparent resistance to considering an alternative possibility that something is seriously wrong that raises the red flags in this case.

Unmemorable Water

With this worrisome example in mind, let's now return to the possibility that water maintains some molecular memory in micro-microdilutions. As we mentioned, for many people this scientific claim had a much more immediate personal implication than the hope of cold fusion, since it could be understood as external evidence supporting the approach of homeopathic medicine. The story of how the scientific journal *Nature* handled the water memory claim is a colorful one, reflecting the quandary presented to *Nature*'s editor, John Maddox, when a reputable scientist, Jacques Benveniste, submitted an article for publication making the extraordinary (fantastical?) claim that a part-in-10^{120} water dilution of an antibody solution still showed evidence of biological activity of the original antibody. On the one hand, Maddox apparently wanted to encourage novel, break-the-mold science that went beyond the current conventions, but in this case the result didn't just go beyond the current establishment mainstreams; it simply wouldn't fit with the bigger picture—our metaphoric raft—of science. As Maddox later said, "Our minds were not so much closed as unready to change our whole view of how science is constructed."[92]

Maddox came up with a compromise to handle this unusual situation. He would accept the paper for publication in *Nature,* given that the referees couldn't find what might be wrong with it (and with a well-respected lab authoring it, it was clearly not pseudoscience). But given the risk that it might mislead a large international public—for example, about half of the French population has used homeopathy—he added an editorial warning: "There are good and particular reasons why prudent people should, for the time being, suspend judgment."[93] Additionally, he insisted that *Nature* send a team to Benveniste's lab to oversee the rerunning of the experiment. Since it is so difficult to see how we fool ourselves or are fooled by others, Maddox chose for his team Walter Stewart, a physicist from the US National Institutes of Health experienced in debunking suspicious science, and . . . James Randi, "the Amazing Randi," a stage magician who had been showing how apparent paranormal effects were being staged by claimed psychics, like Uri Geller.

Reports describe what sounds like a rather diverting scene in Benveniste's lab at the beginning of this rerunning of the experiment, with Randi occasionally entertaining everybody by performing small magic tricks. First, the visiting team watched the experiment run several times the way the lab was used to running it, with all the different chemical vials identified to the experimenters. Then, as had been agreed beforehand, they reran the experiment several times with randomization and hidden labels ("blinded") for the vials. Apparently, Randi wrapped the hidden labels and taped the package to the ceiling (presumably the ceiling was used for dramatic effect) until the experiments were concluded and everybody was ready to reveal which vials were which. When the results were thus revealed, it turned out that the "blinded" experiment hadn't worked: only relatively undiluted solutions were biologically active, not the extreme dilutions that had generated the excitement. The implication was that some aspect of the experimenters' process had been spuriously creating the apparent result when the experimenters weren't "blind" and knew which vial was supposed to give which result. As reported by Maddox, Randi, and Stewart in the next issue of *Nature,* "We conclude that there is no substantial basis for the claim that [antibody] anti-IgE at high dilution (by factors as great as 10^{120}) retains its biological effectiveness, and that the hypothesis that water can be imprinted with the memory of past solutes is as unnecessary as it is fanciful."[94]

In this report, Maddox, Randi, and Stewart end up pointing out a number of problems with Benveniste's lab's work that ring warning bells with our Langmuir list. For example, apparently the experiment was known to "not work" sometimes (Langmuir #5), and the Benveniste team noted that there were some periods of time when no experiments worked. They had speculations about what might be going wrong with the water used for dilution. They also had internal lab "folklore" about how transferring a dilution from one test tube to another would spoil the effect, and about similar nonresults if one diluted multiple times by factors of three or seven each time rather than a factor of ten each time. The Maddox report also noted that the measurement of the effect was a demanding job (counting one kind of blood cell amid many others) that some lab members were known to be particularly good at (perhaps Langmuir #2—the effect is barely detectible?). And finally, the accuracy of the counting appeared to be better than theoretically possible (Langmuir #3). In particular, two measurements of identical samples agreed too well, and didn't show the measurement error always present in such counting-based experiments.

Clearly the surprise of the original paper already showed Langmuir #1: the effect was completely unrelated to the cause, as subsequent dilutions of dilutions still were biologically reactive. And the idea of the molecular organization of water keeping a memory of one of the myriad substances that had once been present sure sounds like a fantastic theory contrary to experience (Langmuir #4). This leaves us with a rather strong sense that this result is a case of pathological science. As the Maddox, Randi, and Stewart report put it, "We believe the laboratory has fostered and then cherished a delusion about the interpretation of its data." Benveniste never changed his mind.

If, at the time, you had hoped that there was finally, after more than two hundred years, some scientific evidence backing up homeopathy claims, you were certainly disappointed. Of course, readers of this book will be interested that the most direct way to address these claims, the randomized controlled trials of chapter 3, have all discredited homeopathy.[95] In fact, there is a long history of such tests since, historically, one of the first-ever double-blind, randomized controlled trials was an 1835 trial of homeopathy![96]

The Torque-Schizophrenia Hypothesis

Since we don't want to leave readers with the impression that scientists who "swing for the fences" (like Babe Ruth) are necessarily engaging in pathological science, it is instructive to consider a case that *looks* like pathological science, and show why it sidestepped that fate. In his 1977 Presidential Address to the American Psychological Association, Theodore Blau, a highly respected clinician, used the opportunity to put forth a truly bold hypothesis: *We can forecast a risk of developing schizophrenia by asking children to draw circles around the letter X!*

Blau argued (and presented some evidence) that children who draw counterclockwise circles (a tendency he called torque) have a substantially higher risk of schizophrenia. This isn't as crazy as it may sound. Blau argued that his simple test might reveal a problem of "mixed cerebral dominance" interfering with the communication signals between the left and right hemispheres of the brain. But within a decade, numerous studies failed to support Blau's hypothesis.

Why wasn't this another "cold fusion" or "memory water" situation? We think Blau's case is different because he was so forthcoming about ways he might be wrong. Blau noted that "there are serious methodological and base-rate problems inherent in the present study which must be explored and accounted for if this research is to serve as a contribution to preventive activity with children at risk," and he suggested other lines of evidence that should be investigated to test his idea. Perhaps his caution explains why his critics were professional and respectful in their treatment of Blau's hypothesis (although one couldn't resist poking gentle fun with the subtitle "Spinning in the wrong direction?").

WHY DO YOU CARE?

Maybe just a few of us are going to be spending much of our time performing science experiments and then worrying that we might fall into pathological science. But all of us regularly have to make decisions that we would like to be informed by our best understanding of reality (like whether to take a homeopathic preparation). So we do want ways to recognize science gone wrong—and to be able to see the experts that we count on using these, and similar, criteria to test the evidence on which they base their recommendations. This is part of the reason why we have presented two dramatic examples of established scientists sliding into the Langmuir list of pathological science indicators.

But perhaps there is an even more fundamental problem that our case histories exemplify. Whether or not we are actually performing scientific studies, we are *all* subject to falling in love with some beliefs about what must be true and then sticking to them no matter how much they conflict with the rest of the things we think we know. ("Surprisingly, I'm most productive when I'm overloaded and multitasking.") We also make bad excuses whenever these beliefs don't work and, as we began discussing in the previous chapter, notice only the best examples of where our beliefs seem to work really well. ("Of course, that multitasking advantage didn't work on Friday because I hadn't had a good night's sleep!") Even established scientists with well-confirmed discoveries to their names, whose training should make them more resistant to this mental failure mode, can clearly fall into this trap, as our two case studies show. Whenever you notice this happening to yourself, you should stop and think of "water with memory" and "cold fusion," and then back away and say, "I've got to be more skeptical here."

Zooming back out to the bigger picture of "science gone wrong," we will also need to be aware of the frequent use of scientific language to cloak fictions, a la pseudoscience, as well as the existence of scientific claims based on truly fraudulent evidence. On the positive side, we can be more forgiving of genuine, honest science that happens to get a wrong result due to statistical fluctuations of random noise, and even occasional mistakes caught by other scientists. On the yet darker side, although fraudulent data might seem to be the bottom of the barrel of what you might see when you're looking at scientific results, there is another historical misuse of the science banner that is more malevolent: scientific language and claims have been used to rationalize and motivate discrimination, oppression, and even mass murder.

These egregious misuses can primarily be described as the marshaling of studies of human subpopulations to justify policies that hurt those subpopulations (or favor the group doing the studying). The degree of hurt has ranged from mild discrimination to deep, legally supported prejudice to ghettoization, even to genocide. Some infamous examples include phrenology, the eugenics movement, Dr. Mengele and Nazi science, and the Tuskegee experiment. Clearly, we are discussing here a deep and disturbing element of human behavior and a fundamental failure of cultures and civilizations, not simply a failure of science. But it is important for us to ask how our use of science could fight against such dangers and resist being co-opted to support, or even be complicit in, such oppression.

Throughout this book, we draw attention to ways that we have learned over the years (often through the culture of science) that we are likely to fool ourselves and misunderstand reality—and the oppressive use of science might be seen as an evil edition of this self-deception. We generally can't prevent fooling ourselves again in these ways, but we can (1) raise our own awareness of these failure modes so we have a fighting chance of recognizing when we start to fall into them and correcting course; and, occasionally, (2) implement techniques and guardrails to help us avoid falling into a given failure mode. For the dangers raised by studies of human subgroups, we often use the first of these approaches, raising our awareness: we scare ourselves with the historical stories, so we have our antennae up when we see a policy based on such a study, and are more sensitive to warning signs when we find ourselves involved in creating or using a study that in any way reminds us of these infamous historical precedents.

We can also offer at least a few examples of the second approach, developing guardrails: After the news reporting on the Nazi human experiments, the Tuskegee experiment, and several other bad cases, the US Congress created a National Commission for the Protection of Human Subjects of Biomedical and Behavioral Research.[97] In the "Belmont Report," this commission set forward three principles for ethical scientific study of human subjects: *Respect for Persons,* which includes protecting their autonomy; *Beneficence,* which includes abiding by the philosophy of "Do no harm" while maximizing benefit; and *Justice,* which includes the fair distribution of costs and benefits. These principles became the guardrail criteria that independent Institutional Review Boards (IRBs) use to approve and exercise oversight over proposed studies of human subjects. Today, for example, every US university and research organization that receives federal funding must have a standing IRB that scrutinizes any proposed research that will involve human subjects, even if the research itself is not federally funded. These IRBs initially focused on medical research but were soon extended to social science research. Many other countries have similar procedures, and there are often separate IRBs to scrutinize animal research.

Like all human endeavors, the IRB process is imperfect, and it illustrates the kinds of "false-positive/false-negative" trade-offs we discussed earlier in the book. For example, the IRB can impose a very high standard that is nearly impossible for research proposals to meet, preventing any possible harm to human subjects but also blocking a great deal of research that could potentially contribute to future human welfare.

Another recent effort to create and institute ethical guardrails has been the development of "community-based participatory research," sometimes abbreviated as CBPR. The idea here is to engage members of the studied community as full partners in the research effort. The community members or their representatives participate in all aspects of the research, from identifying the useful research questions, through the research planning and data collection, to the data analysis and interpretation of results. This isn't always practical, but it is particularly sensible to explore for topics that are controversial and have high stakes for the community. The goal of this approach is to ensure that the community benefits from the knowledge and policies that come out of the research, ideally ensuring that at least these studies of human subpopulations serve the subpopulation well.

These attempts to check the possible misuse of science by humans studying humans are valuable, but there are no simple solutions. We still need to struggle over all the difficulties of the topic, because there are certainly important reasons for us to study human subpopulations, even when the community cannot participate or when there is no particular definable community to consult.

Much of this book is devoted to extolling the virtues of scientific thinking and scientific methodology as a means of understanding the world, so that we can make efficacious decisions based on this understanding. But we well know that a "made with science" label on the cover of any argument is not sufficient reason to accept the argument. We hope the reader is now sufficiently on guard for all of these multiple ways in which, when "science" is claimed, you aren't lulled into accepting it without some careful thinking. In the next chapter, as an antidote to these dark reflections, we will turn to a relatively new approach to addressing a key aspect of all these failure modes.

CHAPTER 14

Confirmation Bias and Blind Analysis

Is there an emotional arc to working on any challenging project? When we discussed the idea of scientific optimism, we were thinking about how it helps us stick to a problem amid the many iterations often necessary to make any progress at all. This was a bit of a hint that challenging projects generally go through some pretty rough patches, whether or not they are ultimately big successes. The vast human project of learning how to think clearly so we can understand the world well enough to thrive—or at least to make efficacious decisions—certainly has its rough patches too. And at this point in the book, after the past couple of chapters, we might begin to despair or at least have serious doubts about that human ability to think clearly.

It is exactly these moments of possible despair, when we might throw our hands in the air and give up, that call for scientific optimism and its motto, "Despair not!" Our 3MT goal here is to see these failure modes in our thinking as simply another, larger problem for us to iteratively work on, so long as we see any signs of progress. And we certainly have evidence of signs of progress. So many of the thinking approaches that we have been discussing in the first three parts of this book represent dramatic conceptual advances just in the past century. This is not a moment to give up. Instead, we must start diagnosing the problem we have just discovered and looking for possible solutions.

In both the worrisome stories of pathological science and the downright frightening stories of scientific claims in the service of human oppression, the protagonists take the same route to shoring up their assertions. That route is paved with confirmation bias, the human cognitive failing we highlighted at the end of chapter 12 as being particularly pernicious.

Let's take a few moments to consider just how pernicious confirmation bias is. As we noted in chapter 12, people often begin testing a belief by looking for facts that support it. That just seems logical. A famous puzzle, the Wason four-card problem, shows this off quite well. The following four cards with a letter on one side and a number on the other are placed on the table in front of you:

A K 2 7

You are asked: Which card or cards do you have to turn over to test the claim, "If there is a vowel on one side of the card, then there is an even number on the other side."

What would your response be? Give it a quick try—we'll wait.

According to the principles of deductive logic, the correct response is to turn over the "A" and "7" cards—turning "A" to make sure there's an even number on the other side, and "7" to make sure there is no vowel on the other side. But generally, less than 5 percent of people make both choices, and about a third only turn over the "A" card. It is correct to turn over the "A" card, which is a way of *confirming* the hypothesis, but people overlook the "7" card, which is a way to *disconfirm* it if it is wrong—another example of a confirmation bias. (Almost half choose to turn over both "A" and "2," but turning over the "2" card won't test the proposition because "If *vowel,* then *even number* on the other side" does not imply, "If *even number,* then *vowel* on the other side.")[98]

Confirmation bias leads us to only seek confirming evidence, while failing to seek disconfirming evidence. But there is a corollary of sorts—a "disconfirmation bias": While people tend not to seek disconfirming data on their own, when they do encounter evidence against their hypothesis, they often subject it to much tougher scrutiny than they applied to congenial evidence. Whereas a simple misunderstanding of deductive logic would constitute a "cold confirmation bias," being tougher on evidence we don't like constitutes a "hot confirmation bias"—a form of motivated reasoning in which we are biased toward making sure our preferred conclusion wins out over an alternative that would show we were wrong.[99]

Rob likes to illustrate this hot bias using his own behavior when he was researching the effects of government policies on drug use. In 1975, Italy decriminalized possession (but not sales) of all psychoactive drugs. A citizen referendum in 1990 recriminalized drug possession. Only three years later, another citizen referendum re-*de*criminalized drug possession. To Rob, this looked like a "natural experiment" that ought to tell us a lot about the effects of drug laws on drug problems. Unfortunately, Italy did not routinely assess drug use in those years, but they did record "drug-related deaths." When he plotted the data, he found that drug-related deaths rose steadily from the mid-1980s. But to his surprise, the death rate dropped off after the 1990 recriminalization. And it started rising again after the 1993 re-decriminalization. Taken at face value, the story seemed clear: penalizing drug possession saves lives. Why was Rob surprised? Because in 1993 he'd published a comprehensive analysis arguing that decriminalizing possession has almost no detectable effects on drug use and offering various theoretical reasons why that was so.

The data from Italy threatened to show that Rob's earlier conclusion was seriously in error, and he felt threatened by that, so he started digging deeper, and found two other European countries that also reported drug-related deaths over the same period. One of them, Spain, also decriminalized drug possession at about the same time as Italy, but never recriminalized it. But despite these different legal histories, the data showed that all three nations saw a temporary bump in drug deaths around 1991, demolishing the hypothesis that decriminalization leads to a higher death rate. Rob was relieved; his 1993 arguments still held.

Was Rob's conduct good scientific practice? In hindsight, he argues (with much embarrassment), the story shows a cold confirmation bias in action, in which he looked for evidence to support his expectations without seeking evidence that might contradict them; but also a hot bias, in which he introduced new data and a more thorough test only when he didn't like the initial results.

You might think there would be less "hot bias" motivation in the more serene world of physics measurements, or, more remote still, cosmology measurements. After all, it doesn't really change our political goals or our financial position if the universe's expansion rate turns out to be 50 km/sec per megaparsec (Mpc) or 100 km/sec/Mpc.[100] But this is precisely the debate that raged for almost twenty years, starting in the mid-1970s. One team of scientists published paper after paper that found the 50 km/sec/Mpc number, while another team published paper after paper that found the 100 km/sec/Mpc number. Both teams were comprised of excellent scientists, and both made good arguments and presented strong data in their papers that led to their results, but it is fair to say that during this period you could pretty much guess who the authors of a given paper were by its conclusions. What was going wrong?

Similarly odd trends were recently discovered in the history we discussed in chapter 5 of particle physics measurements that were repeated over many decades with steadily improving techniques. The oddities generally appeared as measurements that tended to agree with previous measurements better than the measurement noise should have made possible. In this case, too, excellent scientists were doing the measuring, and their results are definitely not understood to be fraudulent, nor the product of pathological science. Again, we may ask, what's going wrong?

Eventually, the likely cause of both of these examples was identified as another insidious effect of inadvertent confirmation bias. This particular source of confirmation bias may not have been as common back in the days when a scientific analysis might consist of making twelve measurements, say, adding them together, and dividing by twelve to get the average. But today, physics measurements are much more complex, usually involving the collection of huge amounts of data, feeding it all into a computer, and writing lots of computer programs to help you draw conclusions from the data. You then spend a lot of time selecting the trustworthy data and debugging your computer programs. For all scientists, these are important jobs. A large dataset will usually include data that was collected incorrectly—perhaps the measuring device wasn't warmed up or a study of a bunch of adults accidentally got some results from teenagers—and these data need to be identified and removed. And it turns out that any reasonably complex computer program is bound to have a number of difficult-to-find bugs. Some of these won't affect the results badly, but some will—and they have to be found and fixed.

The problem is that there is a real tendency to find and reject untrustworthy data, and to find and fix software bugs, right up to the point where you get an answer that doesn't surprise you. That is, you hunt for bad data or bugs when you get results that are unexpected, but you stop hunting when the results "look right"—even if these nice-looking results are produced by some remaining bad data or undiscovered bugs in the computer program. The consequence is that the final measurements and the science papers that result will tend to show results that are systematically biased toward what the scientists expected to see. This looks like the explanation for why the historical physics measurements didn't bounce around as much as they should have based on the random noise expected for those measurements. And in the more dramatic case of the measurements of the expansion of the universe, it is likely that the difference between the 50 and 100 km/sec/Mpc results was strongly affected by the choices each team made of which data was considered trustworthy.

It is somewhat surprising that this particular confirmation bias failure mode hasn't done worse damage. In these historical examples, we can see the saving grace of other aspects of science at work, since eventually the measurements improved enough that these particular bad results became obvious. But in both cases, we lost years of research/discovery time to these biased results. And the danger of making decisions based on such confirmation-biased measurements still exists when we don't have the advantage of hindsight after years of improved measurements.

One of this book's repeated themes is that we can learn from our experience in the science world, which often shows us our failure modes in rational thinking, and carry the lessons over to the rest of our day-to-day life. In this case, it's pretty clear that we are talking about a failing that we all have experience with. Who among us hasn't asked for advice from one experienced person, and then, when we didn't like the answer, asked more people until somebody we trusted gave us the advice we wanted to hear? At this point, we probably stopped asking further advisors, just like the scientist stopping the debugging when they saw the results they expected.

Our goal in finding the examples from the science world, though, is to be able to take advantage of the tricks and tips and techniques that science has invented to help us with these mental failings. In this case, once physicists started to recognize the pervasiveness of this newly identified problem, they began inventing techniques to avoid it—techniques that we should all know. This part of science methodology is new enough that it has not even spread throughout all the different fields of science yet, so readers of this book have a chance to get ahead of some of the science world in this respect, at least for a few years.

AN IGNORANCE-IS-BLISS SOLUTION

At first glance, these confirmation bias failures seem hard to address. It's not like we can expect people to keep working on a problem forever, hunting for bugs in their computer programs, or asking more and more advisors, even after they get the answer they want to see. The core of one of the more effective solutions turns out to come from a surprising place, a "place of ignorance." One of the more fun decision-making techniques we learn as young children is the cake-sharing strategy. When we wanted to share a piece of cake fairly, we learned the algorithm, "You cut and I choose." Since the person who cuts the cake doesn't know which piece they will get, the incentive is to be scrupulously fair in the cutting. We are making use of this ignorance to get the outcome right.

The same idea can be brought to bear on the confirmation bias problem. Imagine that you are the scientist hunting down and rejecting bad data and debugging your computer program, but you don't know whether the bad data and computer bugs you are discovering "fix" surprising results you are seeing. In fact, suppose you don't let yourself know what the final effect is of any of your data cleaning and debugging until you have explicitly decided that you are done with the analysis and are ready to publish—whether or not the results are what you expected, and whether or not you like the results.

Effectively, you now don't know which "half of the cake" will be yours—whether you will be pleased or disappointed—when you make the decision about, say, what data to reject or when to stop hunting for bugs. This ignorance forces you to be honest with yourself about all these small decisions that you must make when completing an analysis, so those decisions are never confirmation biased by the results that you would like to see or expect to see. Physicists call this science innovation blind analysis, in analogy to "double blind" experiments, where the doctors and patients don't know who is getting treatment and who is getting a placebo.[101]

Why doesn't every scientist use this procedure immediately after hearing about it? Blind analysis does take a little getting used to, because most scientists have come to depend on looking at the properties of the results to figure out if they have a problem with their data or their computer program. It requires us to invent new ways to hunt for such problems without revealing the key results, and this takes some retraining and sometimes creativity. For example, imagine a scientific study where the key result does depend on the average of a large set of measurements. A simple approach to blinding here is to hide the average value from yourself by adding a single number, kept secret by a friend, to every data point before starting the analysis. Then, after you have decided you have finished with your final analysis, your friend reveals the hidden number so you can subtract it and find out that average value. In this way, you can debug your experiment by, e.g., scrutinizing data points far from the (hidden) average, and potentially rejecting them as bad (perhaps the detector was broken that day?), without prematurely revealing that key final answer.

This approach to blind analysis is very similar to the one that Saul's group started using when, several years after that debate about the current expansion rate of the universe, they were studying the next cosmology question: How much has that expansion rate *changed* over the eons? Saul's team learned of the then-new blind analysis approach, and, given the experience of those earlier teams debating the related cosmology measurement, it seemed worth a try. In the end, blind analysis became a standard for many cosmology measurements. Originally it seemed like a lot of extra work, but it became even more apparent that it was necessary work when Saul's group reanalyzed other teams' earlier measurements and found clear evidence that bugs had been tracked down just to the point that the results were the expected ones—and when, worse, his team reanalyzed some of their own earlier data and found a similar case! Despite how careful he thought his team (and presumably the others) had been, this form of confirmation bias had snuck in. Once you recognize how insidious this confirmation bias can be, it becomes hard to feel confidence in any result that hasn't used *some* method like blind analysis to avoid such a common problem.

SURPRISE, SURPRISE!

The gamelike aspect of hiding something from yourself leads to some dramatic stories once entire research teams are using this approach. After forty years of iterative development, the international team hunting for gravitational waves was getting ready for the first running of an instrument sensitive enough to detect these almost inconceivably small signals. The Laser Interferometer Gravitational-Wave Observatory (LIGO) collaboration was worried that they would either miss real signals (expected to be quite rare) or falsely identify noise as a gravitational wave (since it is expected to look almost like the real signals, *and* it is overwhelmingly more frequent so will constantly be causing the detector to "cry wolf"). The huge team therefore decided to play the following game with itself. They set up a small subteam that was responsible for occasionally putting a signal into their system that looked just like a gravitational wave, sometimes very weak, sometimes stronger, as a test to see that the rest of the team would take each of these rare signals seriously. In fact, the subteam wouldn't reveal whether a signal was real or not until after the rest of the team had completed the analysis, written the science paper, and, if the signal was real, committed to publishing it.

The first time Saul heard about this system, the large team had been working for months on the analysis and writing the paper, and then, in the end, the subteam revealed that the signal they were examining for the paper was a fake. Several years later, the team detected a signal they believed was obviously fake—the signal was strong, much stronger than needed for their detector to work. But this time, to their great astonishment, the signal turned out to be real, the first gravitational wave detection ever!

For yet another example of an approach to blind analysis, consider the case in which you are testing the effectiveness of different doses of a biological treatment by comparing treatments, each with an increasing dose of some substance. You could have your friend randomize the order of your test tubes and only tell you which is which after you have completed all your analyses. This example should sound familiar, since it is essentially what we saw the Amazing Randi do to blind the test of water with memory. And of course, we wish the scientists studying cold fusion had known of such mental hygiene techniques.

DO TRY THIS AT HOME

If blind analysis were only good for science experiments, it would already be important, but the approach is also useful in daily life. As a particularly simple example, if you really want to know which wine tastes best to you regardless of price and label, you could do a blind wine tasting—which of course is what the wine experts do.

One of our favorite examples of blinding in real life is offered by Annie Duke, a psychologist and former professional poker player, in her book *Thinking in Bets: Making Smarter Decisions When You Don't Have All the Facts.* Duke explains that professional poker players often omit the outcome of a hand when discussing good and bad plays with colleagues: "Telling someone how a story ends encourages them to be resulters, to interpret the details to fit that outcome.... If the outcome is known, it will bias the assessment of the decision quality to align with the outcome quality."

Duke admits that her nonexpert audiences are often frustrated when she discusses strong and weak poker plays without telling them the outcome of the play. When they ask, she explains, "It doesn't matter." In poker, as in life, particular outcomes are influenced by our strategies but also by chance factors we can't control. Good poker players recognize that smart strategies improve our odds over the long run, even when they don't succeed in any single instance.

Blinding is increasingly used by conscientious employers to reduce discrimination against underrepresented groups in hiring decisions. The most famous example is the use of a screen to hide the physical appearance of musicians when they audition for professional orchestra positions, a method that has significantly increased the selection of female musicians. Increasingly, especially in Europe, employers redact demographic information from job applications before screening candidates for interviews.[102]

Blinding is not a panacea, and it can have unintended consequences. For example, many jurisdictions in the US have adopted a "ban the box" policy that prevents employers from including a "criminal history" checkbox on job applications, but several studies suggest that this can lead employers to use minority status as a proxy for (blinded) criminal history, resulting in an *increase* in discrimination.[103]

A P-HACKING GAME TO (CONFIRMATION) BIAS YOUR POLITICAL BELIEFS

You may remember our chapter 7 discussion of the look elsewhere effect, where, if you look at some data in enough different ways, with different analyses, eventually you'll find what appears to be a result that supports a hypothesis you were hoping to prove, but it will really be due to the random fluctuation of noise. That's exactly the kind of danger blind analysis is designed to counteract. And humans are very good at finding post hoc justifications for focusing on the analysis that just happens to support their hypothesis, even if the other analyses don't. Trying out different analyses until you get the result you want is called p-hacking, named by the psychologists Uri Simonsohn, Joseph Simmons, and Leif Nelson.

If you'd like to experience the look elsewhere effect and the tendency to get results you expect when you analyze without blinding the results, you should check out FiveThirtyEight's online interactive p-hacking game (projects.fivethirtyeight.com/p-hacking). It lets you explore an important and controversial question: Are Democrats or Republicans better for the economy? They use a real dataset, too. The trick is, you can define both the independent variable (who is in power) and the dependent variable (how's the economy) in a lot of different ways. Does it matter more who's president, or who's governor? Should "the economy" be a function of GDP, of employment, of inflation, stock prices, or of a combination of all four? Depending on how you define your variables, you can get a statistically significant pattern that makes it look like your party—whichever one that is—is definitely better for the economy.

Once you've found an analysis that shows that your party is better, you can come up with post hoc justifications for why you've defined your variables as you have. Then you might find yourself even more convinced than you were before that your party really *is* better for the economy. The only way to avoid this altogether is to do blind analysis. You could, say, blind yourself to which party is which until you're satisfied with how you've defined the variables. Then when you unblind yourself, you have to stick with that result. Even if it isn't what you wanted. And sometimes, it won't be. Reality only sometimes aligns with our wishes.

BLINDING THE EXPERTS

One of the other most important applications you are likely to find for blind analysis outside of scientific research is in evaluating expert opinions. A dramatic example of this can be seen in the use of forensic experts in crime cases. Apparently, some of our most cherished mystery-story tropes, including the damning fingerprint evidence, turn out to suffer from confirmation bias concerns. Fingerprint experts had a long tradition of professional practices that even quite recently were anything but blind analysis, giving them a strong leaning toward certain fingerprint identifications. As modern scientific practices were brought to the discussion, new studies cast doubt on the calibration of the fingerprint experts, to the point that reports from leading scientific institutions recommended that such legal evidence required fresh calibration work.[104] (Strikingly, multiple respected fingerprint experts changed their conclusions when presented with blinded tests that included fingerprints they had previously identified.)[105] Such blinded calibration work has now begun to be part of the forensics story.

With this in mind, we should expect experts whom we consult to try to give their opinions using blind analysis whenever possible. This will often be a bit of a change in common practice. For example, doctors generally see the conclusions of the previous doctors consulted before giving a second opinion.[106] But if you are the patient, you would likely benefit if the doctor were blinded to earlier conclusions. Similarly, from our perspective, judges and jury members would benefit from expert opinions based on blind analyses, although some disputants might prefer the current unblinded approach.

OPEN SCIENCE

Outcome-blind analysis is only one way of reducing confirmation bias. There are others, and they are increasingly discussed together as part of a broader trend often called "the open science movement."[107] We are excited about this movement, because it is a big part of what we are calling Third Millennium Thinking. While many of the methods endorsed by the open science movement predate it, the strength of the movement is to pull them all together into a general approach toward improving the credibility and integrity of scientific practice.

One such open science method—one that is increasingly required by top academic journals—is *preregistration,* where prior to conducting the study, investigators publicly archive their proposed methodology, specific hypotheses to be tested, and the data analysis plan they will use to test them. Some have argued that preregistration eliminates the need for outcome-blind analysis (or vice versa), but the two approaches are complementary and should be used in tandem. Preregistration forces investigators to think carefully about their methods and analysis strategies before they collect data, while blinded analysis enables the team to test new ideas that hadn't occurred to them before, or handle new analytic problems they hadn't anticipated, without fear of letting their wishes cloud their research.

Open science advocates have championed a host of other practices, and we will just mention a few of them. One is the more routine use of multilab investigations, where multiple labs agree to conduct their own test of a particular hypothesis; this is mostly done in an attempt to replicate provocative findings in the literature but can also be done for important new research questions. A variation is the use of "many analysts" projects, where independent teams are given the same dataset and asked to test the same hypothesis. (These practices triangulate among different labs' and different analysts' confirmation biases with the hope that not all will be biased in the same direction.) Another strategy that is gaining steam is for journals to devote space to *registered reports,* in which the researchers propose a study to the journal, and the journal commits to publishing it before the study has been conducted, as long as the theoretical justification and the proposed methods are approved by technical peer review. In addition to removing the confirmation bias due to journals being less likely to publish a finding that is way out of line with expectations, this practice also removes the natural bias of journals to only publish results that prove interesting hypotheses and neglect those that disprove them. And when two or more investigators strongly disagree about a hypothesis, they can form an *adversarial collaboration*. In this approach, the competitors jointly design, conduct, and publish one or more experiments that both view as fair tests of their competing perspectives. Each researcher reserves the right to state their own conclusions at the end of the paper, but by precommitting to the methodology they can't dismiss unwanted results by attacking the study.[108]

Open science advocates have also lobbied hard to persuade journals to either require some or all of these practices or reward authors for engaging in them by showing icons or "open science badges" crediting authors for each practice they engaged in.[109] They have also championed various ways to encourage scientists to be more willing to openly admit when they think their prior work is mistaken or flawed (the Loss-of-Confidence Project we discussed in chapter 5 is a good example of this).[110] Importantly, researchers are using the tools of science to scrutinize these proposals, and empirical evaluations of the costs and benefits of these methods are beginning to emerge. We hope that the set of open science tools will expand and become more refined over time.

WHAT TO TAKE WITH YOU

Now that you are familiar with blind analysis, you should be looking for it—or one of these alternative methods for reducing confirmation bias—every time you choose an expert or accept an expert opinion. If a trustworthy expert is one who recognizes the ways we fool ourselves, we want to find experts who show that they are reaching their results (and favoring others' results) with appropriate blinding, preregistration, or other safeguards against confirmation bias. And more personally, you should be on guard anytime you find yourself determining some yes/no result, some numerical value, or some choice between actions if you presume to know the answer before you study the problem. You may need to decide whether to start taking statins to lower your "bad cholesterol" level, or whether to pay for advertisements to drum up business. You may be trying to help your college-age child choose a major that will allow them to make a living.

To find the answer to such a question that best reflects reality, you'll undoubtedly need to decide what data to use and how to use it, for instance, which website's data to trust and which to ignore. (You may be tempted to say, e.g., "Let's just use the data from the WebMD site rather than the data from the Mayo Clinic site, because it is easier to read—and looks less scary in terms of health outcomes I am hoping to see.") You will have to decide which data are close enough to your case to be relevant.

(And if you don't like the first results you see, you might then find yourself suggesting, "I want just the medical data that applies to my age group and general level of fitness.") However, you should *not* let yourself make any of those decisions if you're also seeing the consequence of these data choices on the final answer—because then you'll subconsciously cherry-pick information in such a way that you'll get the answers you want. To assess the quality of the site's methodology without seeing its results, you may need to enlist a friend to cut-and-paste each site's information into a separate document, taking out the name of the site and hiding the conclusions.

Henceforth, anytime you're in this kind of decision-making situation, where your selection of facts can be influenced by the decision they would lead to, stop everything and, in a dramatic tone of voice, announce to everyone around you (ideally, they are working on the same problem): "We have to stop doing this study (or thinking about this problem or choice of actions) right now, because we're starting to do things that are going to bias the result. We have to *not* look at the result until we've chosen all of our data and checked all of our decision-procedure plans!"

This is not the last time we will need to invent a new technique to avoid fooling ourselves—and also not the last time we will despair of our ever keeping ourselves thinking clearly. Because every time we invent a new technology that changes the way we interact with the world, with one another, and with our collective study of reality, we come up with all-new ways to fool ourselves. Next time, it won't just be a matter of computers changing the rules of the game by offering us more complex analyses and more opportunities for programming bugs. For example, once we all get implants in our heads, and answers to our thoughts get projected in realistic-looking images on our eyeglasses (now *that's* a scary scenario!), we'll need to invent new techniques to avoid fooling ourselves.

The scientific optimist's life—the 3MT life—is perforce a truly creative life.

Part V

Joining Forces

CHAPTER 15

The Wisdom and Madness of Crowds

So far, we've looked at the rationality of *individual* minds. But of course, most big problems get tackled by groups of individuals working together—couples revising their budgets, neighbors developing an earthquake-preparedness plan, coworkers coordinating a complex project, NASA engineers safely landing a Mars rover, and all the work of city councils, state assemblies, Congress, and the United Nations.

But are group minds anything other than a collection of individual minds? Are groups more—or less—than the sum of their parts? Historically, observers have offered two starkly different views of groups—one pessimistic and the other optimistic. The problem is that neither view is entirely persuasive. The bleak view comes from post hoc descriptions of real-life groups that performed very badly. The rosy view comes from a simple demonstration people have been doing since the early twentieth century that can make groups look brilliant... until you realize that there's no real group discussion involved at all.

So, after introducing you to those two views, we'll take a look at what has been learned from more systematic experimental research on small-group performance, where we can manipulate various features of the group and the task and observe how the group reaches a final decision. That research shows that the pessimistic and the optimistic views are each correct some of the time—making decisions in groups can indeed make things either better or worse. Fortunately, we can identify some of the factors that make group processes go well or go poorly. Understanding the conditions under which groups move to one or the other extreme allows us to begin to develop a recipe for how and when groups work best, a point we begin to address at the end of the chapter and continue to develop in the next two chapters.

MOB BEHAVIOR AND GROUPTHINK

An early pessimistic view of groups is provided by Charles Mackay's 1841 book, *Extraordinary Popular Delusions and the Madness of Crowds,* which introduced the notion of a "herd mentality." Mackay offers lots of case studies of crowd delusions and mass hysteria, including the famous Dutch tulip mania, where people became so convinced of the investment value of tulips that some even sold their houses to raise enough capital to purchase a single bulb. (A more theoretical account covering similar territory is Gustave Le Bon's 1895 book *The Crowd: A Study of the Popular Mind.*)

Most of Mackay's examples involve collective behavior, but they don't involve formal groups making collective decisions (committees, etc.), and in many cases, the people involved don't even know one another. They are the kinds of examples people often label "mob behavior." Some of the most chilling empirical research has examined historical records of lynch mobs terrorizing Black Americans. Statistical analyses show that the probability that these groups killed their victim (usually by hanging) increased with the size of the crowd, and that it was higher when the events occurred at or after dusk.

Crowd size and darkness increase anonymity. Your individual identity counts for less and less the more people there are in the crowd; and the less light there is, the smaller the chance that others will recognize you. So as your own individual identity is reduced and your sense of anonymity increases, there occurs a diffusion of responsibility where no one feels a sense of agency. Psychologists call this *deindividuation.* There is another phenomenon contributing to mob behavior called *emotional contagion,* where exposure to someone else's emotions triggers similar emotions in ourselves. As you express sadness at something, I start to feel it too; you perceive me feeling sad and that amplifies your sadness, in a vicious feedback loop.[111]

Many of the examples of deindividuation and emotional contagion seem extreme and rare. But in fact, less extreme examples occur quite often, as readers may have observed if they've ever attended school board or city council meetings on contentious issues.

An equally grim view of the irrationality that can flourish in groups comes from Irving Janis's 1972 book *Victims of Groupthink: A Psychological Study of Foreign-Policy Decisions and Fiascoes.* But rather than studying disorganized public "mobs," Janis closely examined decision-making by elites—most famously, John F. Kennedy's presidential cabinet during the Bay of Pigs fiasco—a failed attempt to support a coup in Fidel Castro's Cuba—and the later Cuban missile crisis. Janis received harsh criticism from his fellow academic psychologists (who argued that the cases he cited weren't randomly sampled and lacked the rigor of controlled lab experiments) and historians (who questioned the accuracy of the accounts). But the book has had a lasting influence, deservedly so, because Janis had such a knack for identifying—and labeling—group pathologies, and there's plenty of evidence that the ones he describes do sometimes happen in real-life groups.

Although the term was introduced by William Whyte in 1952, Janis offered his own definition of groupthink in *Psychology Today* in 1971, calling it:

> [A] quick and easy way to refer to the mode of thinking that persons engage in when concurrence-seeking becomes so dominant in a cohesive ingroup that it tends to override realistic appraisal of alternative courses of action. Groupthink is a term of the same order as the words in the newspeak vocabulary George Orwell used in his dismaying world of *1984*.

Janis listed eight main symptoms of groupthink: illusions of invulnerability; belief in the inherent morality of the group; collective rationalizations; stereotypes of out-groups; self-censorship; illusions of unanimity; direct pressure on dissenters, and "self-appointed mindguards" (members who protect the leader from exposure to conflicting viewpoints or information). He also proposed two main risk factors that produce these symptoms. The first—a "provocative situational context" involving a looming threat and no obvious solutions for avoiding it—is largely beyond the group's control. The second risk factor is a particular form of unhealthy organizational culture characterized by insulation of the group, a lack of impartial leadership, and a lack of heterogeneity in members' background characteristics and ideologies.

Janis offered a variety of potential solutions to the problem of groupthink. He argued that leaders needed to refrain from voicing their own opinions, and perhaps from attending some meetings, so that members would feel free to speak candidly. Group members should be encouraged to play devil's advocate, critiquing the group's proposed solution and arguing for the merits of alternative solutions. Groups should be split into subgroups to deliberate independently.[112] The groupthink concept was not originally based on, or applied to, the conduct of scientific teams, and it was developed in the study of policy elites rather than ordinary citizens. But it seems readily applicable to scientific teams, PTA committees, university faculty meetings, and really any setting where people need to work together to solve problems.

THE WISDOM OF CROWDS

Okay, so much for the pessimistic view of groups. A much more optimistic view is the "wisdom of crowds" perspective. In 1907, Francis Galton published an article in *Nature,* documenting a remarkable finding. Galton had asked large numbers of people to guess various quantities, e.g., the height or weight of an object. He showed that even when most people's guesses were way off, if you computed the average of all their guesses, you'd find it quite close to the true value. Here's an example: In a class demonstration one year, we asked everyone to estimate the weight (in pounds) of the heaviest living ostrich. We had estimates ranging from 15 pounds to two tons! But the mean of their responses was 326 pounds—very close to the correct weight of 345 pounds. Even though we know (and will soon tell you) exactly why this works, it still feels magical to us. It's as if there's some invisible group mind.

Many real-world examples of this wisdom of crowds effect were documented by Benjamin Page and Robert Shapiro in the 1992 book *The Rational Public,* and by James Surowiecki in the 2004 bestseller *The Wisdom of Crowds.* Both books have a buoyant, celebratory feel, their authors reveling in the idea that while we can each be ignorant in our own way, if we pool our ignorance, we can be brilliant without anyone even changing their mind.

We hate to rain on the parade, but there's nothing magical going on here. And what is going on has nothing to do with group minds, or even human capabilities. Rather, as those authors acknowledged, the wisdom of crowds effect is a purely mechanical result of a fundamental concept in statistics—the "law of large numbers." Recall our earlier distinction between noise (random or statistical error) and bias (systematic error). When we take measurements, or solicit people's estimates, of some quantity, each estimate will contain some random error. The law of large numbers, in its simplest form, says that when we aggregate (e.g., take the average of) a large set of estimates, their random errors will cancel one another out. Consider our students' guesses about ostrich weights. Most people don't know a lot about ostriches, but with a bit of Fermi reasoning (see chapter 11), we can figure out that the right number is going to be heavier than the weights of most people—so probably greater than 200 pounds—and considerably lighter than the weight of a car—so probably well below 1,000 pounds. Beyond that, we're guessing, and because random errors are as likely to be too low as too high, when we aggregate people's responses, the errors start to cancel one another out; e.g., if my guess is off by +50 pounds and your guess is off by -30 pounds, together we are only off by 50–30 = 20 pounds.

Once we understand this effect, we can see it has nothing to do with group deliberation or group intelligence. Indeed, we can demonstrate that allowing group members to interact can worsen performance.

For example, we once asked students to guess the percentage of voters in Alameda County (which includes Berkeley and Oakland) who supported Mitt Romney in the 2012 election. The average estimate was 19.8 percent, quite close to the correct answer of 18.4 percent. But before telling them that, we asked the students to try to harness their collective intelligence by discussing their estimates with one another. The result of that deliberation was a new estimate with a mean of 24.2 percent—in other words, they did worse. This illustrates a dirty little secret of the small-group performance literature, which is that many of the benefits of aggregating opinion can be lost the moment people talk to one another! Of course, in the end, we will be advocating that people do talk to one another, just with careful structuring of the deliberation.

SMALL-GROUP EXPERIMENTS ILLUMINATE WHEN GROUPS SUCCEED AND FAIL

We've argued that the "madness of crowds," "groupthink," and "wisdom of crowds" literatures are not fully credible as a general picture of how decision-making groups typically perform. So now we turn to the experimental small-group literature, in which groups of ordinary people are assembled to complete tasks, solve problems, or make decisions under random assignment to different controlled conditions. We'll generally skip the details of these experiments, focusing on some of the big takeaway points.

A key distinction that helps to organize those findings is between two different kinds of influence in groups. The first is *informational influence* (or "strength in arguments")—the ability to merge the collective knowledge of the group and use it to reason to a solution. The second is *normative influence* (or "strength in numbers")—a tendency to give deference to the largest faction or voting bloc in the group. These two types of influence are sometimes hard to disentangle, because majorities often favor the position with the best evidence and arguments. But in experiments, it is possible to tease them apart by varying groups' faction sizes (while holding evidence constant), or by varying the strength of the arguments (while holding faction sizes constant).[113]

For some types of situations, "strength in numbers"—e.g., a "simple majority wins" rule—seems appropriate. Groups often have to reach decisions for which neither logic nor evidence can provide a "correct" answer. For example, who was the better guitarist, Jimi Hendrix or Andrés Segovia? This really comes down to a matter of taste. One person can argue, "Segovia is a better guitarist because his technique is immaculate, and his études require a level of dexterity that Hendrix couldn't match." Another can counter, "Hendrix is better because his emotional expression and the sheer drama of his playing single-handedly (well, two-handedly) expanded the boundaries of popular music." Thus, groups often decide by majority rule when the topic is largely a matter of opinion or taste—choosing which band to hire for an event, which logo to adopt for a new company, or which candidate to endorse before an election. For very serious issues, it may require a strong majority (e.g., two-thirds), but for many issues, a simple majority may suffice. A majority rule process can bring about closure on a topic for which differences of opinion are unlikely to be resolved.

Still, majority rule can have important drawbacks. Factions can resort to browbeating or bullying to win adherents. Minority faction members can feel resentful, and may be ostracized. And if a group's factions are correlated with personal characteristics like gender or race, some members of the community can expect to consistently be on the "losing" side.

And many problems do have solutions that, once identified, are "demonstrably correct." Once they are clearly articulated, most or all members of the group come to realize that they make sense, so there is "strength in arguments." For this to happen, of course, someone needs to offer the best solution, but that's not sufficient. The group members need to have some kind of shared conceptual system for identifying stronger versus weaker arguments. We'll return to that point in a moment.

Even without observing a group's deliberations, it is often possible to infer something about the balance of normative and informational influence in the group, if you know how the members were divided among positions (i.e., factions) at the beginning of discussion.[114] Thus, social psychologists have found that group decisions on matters of taste or opinion tend to be predicted by which faction was largest at the outset. This is true even when these groups hadn't explicitly agreed that they would be following the majority in making their decision.

How would we detect strong informational influence? At the extreme, a group operating under "strength in arguments" will adopt the best solution or argument that is offered, even if that position only has a single adherent at the start of discussion. Social psychologists call this a truth wins process.

As we noted earlier, for this to occur, there must be some kind of shared conceptual system that makes some positions demonstrably correct, or at least demonstrably superior to the other positions. What do we mean by "demonstrably correct"? It means that the group members have a shared way of assessing a potential answer—a shared conceptual system that can be used to verify whether it is correct.

Arithmetic is one such shared conceptual system. If we are asked, "What's 12 x 321?" most of us probably have no idea off the top of our head. But if one group member shouts out, "It's 3,852!" anyone who knows how to multiply numbers can verify that this is correct.

Another example is a shared understanding of logic. Consider the following: *Jack is looking at Anne, but Anne is looking at George. Jack is married, but George is not. Is a married person looking at an unmarried person? Yes, no, or cannot be determined?* In groups, one of these three answers is quickly selected by most people, but it is fairly easy to see that one of the other answers is actually correct, once it is explained. (We'll let you—or your favorite group—decide.)

Common knowledge can also serve as the shared conceptual system: For example, suppose you're walking with a group, and everyone is lost. You all agree you need to go north, but you disagree about which way that is until one person points to the hills, which everyone knows are in the east.

Science is a domain where we hope to see a truth wins process. Good scientific theories can serve as shared conceptual systems, providing they have a strong internal logic and good support from external evidence. So in science, does "truth win"? We think the answer is "Eventually," but the process of getting there can be slow, and there's no question that scientific majorities can temporarily prevail even when they are wrong. (Arguably, all scientific majorities are still partially wrong—just less wrong than their predecessors.) When told of a Nazi-era book entitled *100 Authors Against Einstein,* Einstein quipped: "Why one hundred? If I were wrong, one would be enough." For Einstein, it seemed obvious that truth will win when we each stop to "do the math." The Einstein example didn't involve simple arithmetic, of course; it involved physics calculations about the behavior of light. Showing that a physics calculation is correct or incorrect is only possible among people who have a shared conceptual system for solving the problem (in this case, a raft of empirically well-tested physics concepts and supporting mathematics). If the three authors of this book had to solve a physics problem, John and Rob would probably take a coffee break and let Saul solve it.[115] Many small-group experiments involve memory tasks, math problems, and logic puzzles where it is usually possible to convince the most people that one particular answer is correct, and as a result, the groups tend to outperform their *average* member. But the process often looks more like "truth-supported wins" than "truth wins," in that it often takes at least two or three members to advocate for a solution before everyone else stops and thinks it through hard enough to see why it is correct. As a result, groups don't always outperform their *best* members. And groups perform a lot worse when they don't have a shared system like logic or arithmetic or scientific reasoning to help them see why one proposal is more likely to be right.

There is another type of situation in which "strength in numbers" often plays an unfortunate role. In an earlier chapter, we saw that individuals are plagued by a variety of systematic biases in judgment (the availability heuristic, the hindsight bias, and so on). It turns out that groups can either amplify or attenuate judgmental biases.[116] When a group is primarily operating under a "strength in numbers" (majority wins) process, the group may actually *amplify* any shared biases in individual judgment, so groups can make things worse. Note that this conclusion does not require the kind of institutional pathologies that Janis saw as risk factors for groupthink—a lack of impartial leadership, insulation from outside voices, etc. If you bring a bunch of strangers together and give them a task that evokes the hindsight bias or the availability heuristic, so long as they engage in a "strength in numbers" process, bias amplification is just baked in.

Can groups ever attenuate biases? Fortunately, the answer is yes, under two types of conditions.

The first is when the biases differ across members. We have seen how random errors get canceled out in the aggregate. Biases, by definition, are not random, so we might not expect them to cancel out, but the key here is whether each member has the same bias, or whether each member has a different bias (e.g., political ideology). If the biases differ, they can cancel out. This is one reason why member diversity can promote better performance.[117]

Note that we say that diversity among group members "*can* promote" rather than "*will* promote" more effective performance. Whether it does so may depend on whether the diversity involves an attribute that helps the group find new ideas for solving the task. And to benefit from diversity, the group needs a culture of respect and participation that allows minority viewpoints to be aired and taken seriously. But of course, diversity can offer other benefits (fairness, legitimacy, novelty, fun) irrespective of its effect on performance.

The second alternative condition that allows groups to overcome shared biases that initially distort their judgment is working under a "truth wins" scheme. Consider, for example, the anchoring heuristic we described in chapter 12: If a group is trying to estimate how much it will cost to renovate a building, their initial estimates can easily be skewed by a salient anchor, e.g., the chair's suggestion that they start with an arbitrary estimate of $100,000 and adjust from there. But if someone says, "Wait a minute, $100,000 is not even close," and proceeds to show that the likely expenses could easily add up to a far larger number, the anchor will diminish in importance, so long as the other members find the calculations plausible.

HOW CAN WE GET THE BEST FROM GROUPS?

As we've seen, there are ample grounds for both optimism and pessimism about whether group decision-making can correct for the biases and irrationality that can lead individuals astray. Fortunately, deciding whether groups will improve on individual decisions isn't just a crapshoot. We can summarize the kinds of conditions that foster optimal group decision-making, and we can strive to implement those in the groups we participate in.

- When possible, group leaders who want to find good solutions should refrain from stating an early preference. Groups can also avoid premature commitment by postponing any votes until there has been a thorough discussion of the evidence.
- Groups should foster a culture of respectful argumentation in which people hear one another out and no one censors their views out of fear of being bullied by the majority. This culture may be encouraged by asking some participants to explicitly play the role of devil's advocate as a way of normalizing the consideration of a nonmajority position.
- Group discussion can feel easier when group members have similar backgrounds, similar biases, and similar tastes. But homogeneity is a lousy recipe for effective decision-making. Diverse groups are more likely to reduce random noise, overcome initial biases, and find more effective solutions, even if it takes more work to do so. Diverse groups have an added benefit of enhancing the group's legitimacy in the eyes of the larger community.
- Groups can reduce the power of judgmental biases and heuristics, but they are most likely to do so when they have shared conceptual tools for recognizing good answers once they hear them. We very much see the material in our book as a set of conceptual tools that can help people do just that, and enable them to build on those good ideas and make them better.

Let's focus a little further on that last point: The challenge here is to develop tools to make it more likely that a group recognizes good answers. What do we mean by good answers? Well, first of all, we want our group decisions to be effective in the world as it actually exists, so it is obviously important that the group is able to recognize, to the extent possible, what is true about the world. Their group process needs to track that shared reality that we discussed in chapter 2. This is certainly easier if the whole group has learned about the concepts we have been discussing throughout the book, such as probabilistic thinking, causal reasoning, and, ideally, even "orders of understanding" and Fermi problems, because those concepts are embedded in a whole system of ways of testing whether hypotheses are supported by observable, verifiable, and reproducible evidence.

But there is another aspect to group decision-making that makes it more complicated than individual decision-making, because decisions are generally not simply determined by getting the facts about the world right. Most decisions depend crucially on our values, as well as the emotions that drive us to make those decisions, whether those emotions are rooted in our fears and desires or our ambitions and goals. In fact, values and emotions are such important drivers of decisions that without a very principled plan for how a group will approach the process of decision-making, it is likely that they, not all of the rational processes of reality tracking that we have been discussing throughout this book, will determine the outcome. So in the next chapter, we turn to considering various principled approaches to group decision-making that account for the values that motivate and matter to individual human beings—spiritual, moral, philosophical, political, emotional, relational—with the ambitious goal of preserving our capacity to think and act rationally. And in the chapter after that, we will look at some new innovations that appear to help group deliberations work more effectively.

CHAPTER 16

Weaving Facts and Values

Imagine you are the mayor of a large European city, and your experts recommend that the city test a new program designed to reduce the criminality and overdose risks of heroin addicts by allowing them to receive heroin for free from medical providers. Should you approve the test? If the test achieves those two outcomes, will you approve the program? How might facts and values enter into this decision? What if the facts point one way (e.g., crime and overdose rates go down) but our values point another (supplying heroin to addicts feels wrong)? And what happens when there are disagreements about values?

For most important decisions we make, ascertaining relevant facts is necessary, but insufficient. Facts don't tell us what to do once we have them. Even if the facts are established to everyone's satisfaction, our decisions about how to act require a consideration of our values and feelings. As we've argued throughout this book, there are productive ways to debate facts—but is it possible to constructively debate value conflicts? And when there are passionate disagreements over values, can it be nonetheless possible to reach a consensus over what to do?

FACTS VERSUS VALUES

We start with the rough idea that there's a difference between, on the one hand, factual or descriptive propositions, such as:

- The earth is 4.543 billion years old.
- Grand Central Station is on Forty-Second Street.
- Flu is a virus.

and, on the other hand, normative or evaluative propositions, such as:

- I shouldn't take this offer.
- It's not okay to say you're going to do it and then back out.
- The government should fix the potholes.

It would seem appropriate to divide up the process of decision-making into, first, determining which values and goals matter to you most in the decision, and second, finding out the objective facts about how the different policy options will affect these values and goals. So you might set up a study in which (a) the social values relevant to determining the outcomes that constituents want would be established, and (b) experts would estimate the program's effect on each of those outcomes. You might begin with a list of features you think will matter to people, and then go over this with community groups, pressure groups, and the general public, to get to a list of the things that are important in considering, for example, drug policy. In the case of the medically provided heroin program, we list three outcomes—the effects on overdoses, criminality, and the number of people using heroin—but there are likely to be many others (the cost of the program, savings in emergency room and law enforcement costs, etc.). We might even consider factors that may be a bit less central, such as the impact of drug policy on tourism. These may be given less weight.

So a possible diagram of how one might approach the decision might look like the one below.

The way we might work out the overall acceptability of any one package of policy changes—in this case, the proposed policy change was to provide medically supervised heroin—might then run something like this. We can ask our experts to determine how much a particular policy package would affect each outcome factor (such as overdose rate, crime rate, and so on; the factors in the middle column) that matters to the acceptability of the proposal. For example, they can look at how much the proposed medically

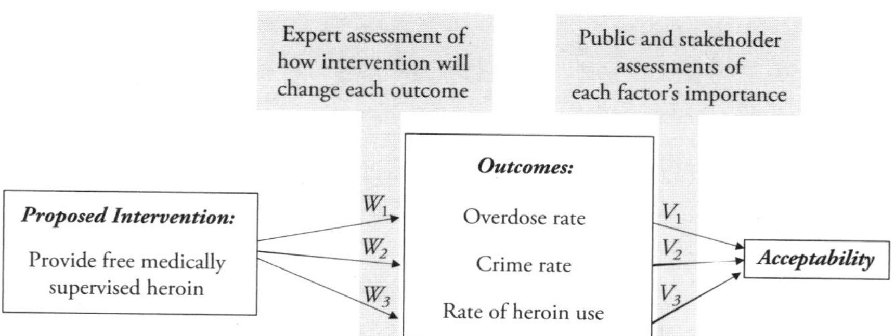

provided heroin policy would affect crime rates, and they can then put a number, W, on how much the policy affects the outcome factor. The more the impact, the bigger the W number.

Meanwhile, we can ask the public and stakeholders to give each outcome factor a number, V, indicating the value that people have given that outcome factor. The more the factor matters to the acceptability of the proposal, the bigger the V number. Now the integration of fact and value is a matter of simple arithmetic. We multiply the W by the V for each outcome factor and add them all up to find the overall acceptability of a particular policy package:

$$\text{Acceptability} = (W_1 \times V_1) + (W_2 \times V_2) + (W_3 \times V_3) + \ldots.$$

If we compare this acceptability number to those of other proposed policies—or the status quo policy—we can then choose the policy package with the highest overall acceptability rating.

So we don't need to let the facts and values just jumble around in our heads before making a decision. We can, instead, assign some numbers and do simple arithmetic. One great advantage of this way of doing things is that it makes a rational review of one's position possible. You might, for example, find that you are in agreement with your opponent about a policy decision on many points but you assign different weights to keeping people from overdosing. So now you can focus the discussion. Perhaps you can convince them, perhaps they can convince you about the priorities.

Still, you might say, this method of decision-making is a bit complicated. It wouldn't work in the real world. When people care passionately about an issue, they're not going to pause for this kind of exercise. They are going to slug it out, trying to impose their preferred solutions on one another.

What we'll now see is that this method actually *did* work in the real world in a particularly fraught situation. It didn't work perfectly—far from it—but for a first attempt, it worked better than one might have expected.

THE DENVER BULLET STUDY

In Denver, back in 1974, the police department made the decision to move from using regular bullets to using hollow-nosed bullets.[118] This was an extremely controversial decision. People said that hollow-nosed bullets were just dumdum bullets, which were outlawed. Dumdum bullets, which flatten on impact, cause severe injuries. There were protests from the ACLU and various activist and community groups. People were outraged at the police department.

At this point, a police officer was killed by a hollow-nosed bullet. Hundreds of police converged on City Hall. They didn't want to be outgunned by criminals using hollow-nosed bullets. Things were pretty fraught and the argument went on and on. One side presented the case for having these bullets, and the other side, the case against having these bullets. It was like the procedure in a court of law.

Experts were called on both sides, but as the dispute went on, it became clear that it was hopeless. There was a general feeling that the whole system of communal law and order was breaking down. Members of the city council and others seemed convinced that the usual adversarial method had failed, and they faced a dangerous impasse.

What was going on was that legislators were asking ballistic experts to tell them what kind of bullet was "best," but because the criterion of "best" couldn't be defined in technical terms, this was a question their expertise, in and of itself, did not qualify them to answer. Expert knowledge was relevant in considering the question, but answering the question involved making value judgments particularly around whose safety to prioritize. And the ballistic experts were not especially qualified, any more than you or I, to make those value judgments.

At the same time, legislators were advocating for or against hollow-nosed bullets without knowing enough about the technical characteristics of bullets or the different types of bullets available to engage in a competent debate. Their opinions were based entirely on value judgments, without knowledge of facts essential to making those judgments wisely. The problem was this: fact and value had not been separated. That was a significant factor in the breakdown of the usual methods of civil government.

As the debate continued furiously and fruitlessly, with law and order eroding and hundreds of police officers massed outside the city chambers, who would you have called in to break the impasse and help achieve some resolution? The council worked with a team of experts assembled by Kenneth Hammond, a psychology professor at the University of Colorado.

Hammond and his colleagues analyzed the issue. They saw that the bullets had characteristics that really mattered in the debate. They distinguished two factors: the degree of injury caused to the person who gets the bullet, and the stopping effectiveness of the bullet. Injury was measured as the likelihood that the person hit by the bullet would die within two weeks. Stopping effectiveness was defined as the likelihood of a person hit by the bullet being able to return the fire. These had not been separated in the arguments previously, but they're different factors. And one other criterion mattered too: the threat to bystanders. How likely was the bullet to ricochet and hit bystanders? It may sound a bit heartless or cold-blooded to make these distinctions. But they matter practically. Everything has to be made explicit if we're going to have any hope of reaching consensus through conversation on hard topics. Every bullet on the market had certain technical characteristics, like its weight, how fast it comes out of the gun, the amount of kinetic energy it loses as it hits the target. Each of these characteristics had an impact on degree of injury, stopping effectiveness, and threat to bystanders, and since the characteristics of bullets differed, so too did their effects in these three areas. These were matters of fact that could be determined by the ballistics experts.

What also needed to be determined were the matters of value. What did all the stakeholders want in terms of degree of injury, stopping effectiveness, and threat to bystanders?

So in the middle of all this uproar, Hammond's team took members of the city council, the community groups, the advocacy organizations, and random members of the general public and sat them in front of computer consoles that described hypothetical bullets in terms of their stopping effectiveness, severity of injury, and threat to bystanders. For each bullet, they were asked to rate the level of acceptability. Then the scientists showed the raters the curve of their results. As the level of injury caused by the bullet went up, the acceptability went down. As stopping effectiveness went up, acceptability went up. As threat to bystanders went up, acceptability went down. So they extracted weights on all these values from everyone. It was an organized route to discovering these values' weights, which included the important emotional reactions regarding human injury.

Once they reached this point, they asked their ballistics experts to review the base technical characteristics of all the bullets available on the market—muzzle velocity; the loss of kinetic energy as it entered the body; the type, size, and shape of the wound inflicted; and how deep the bullet went into the body—and to assess the implications for injury, stopping effectiveness, and threat to bystanders for each one.

At that point, they discovered that there was one bullet they hadn't previously considered. It was a hollow-point bullet, but it was different from the one they had been considering, and it was found by the ballistics experts to meet the criteria for acceptability established by the community members, better than any of the bullets they'd been arguing about. It had good stopping effectiveness, which made the police feel safer, a relatively low degree of injury, and a low risk to bystanders.

Everybody recognized that this procedure hadn't been clearly slanted in favor of anybody—not in favor of the police, and not in favor of the community groups advocating against the use of "dumdum" bullets by the police force. They all agreed that police should use the particular bullet they'd identified in their deliberations.

In fact, the bullet was used for years, without drawing further controversy. The important thing here is that there was a separation of fact and values, which allowed judgments to be made by those best qualified to make them. The decision process had involved a blind study, in a sense, because the experts didn't know what weights community members were putting on the relevant factors, so they didn't know what bullets their expert assessments were favoring, and the people sitting at the consoles making their evaluations didn't know what the experts were concluding about various bullets. It didn't matter if you initially had been on the side of the police or on the side of the local activists, because you didn't know what the outcome of your judgments would be. Another advantage of the procedure was that it was fully explicit. The reasoning could be reviewed and explicitly evaluated.

WHY DON'T WE DO THIS MORE OFTEN?

Many policy disputes would benefit from this kind of approach. One advantage is that it helps to mitigate the problem of "badging" we discussed earlier: that being for or against controls on gun ownership, for example, signals your membership in one social group or another. Hence, when asked directly where they stand on gun ownership, people may simply give the answer demanded by their group identity. But it's a different matter if you are asked what weights you would put on various factors that might be affected by gun ownership policies. There, you really could hope for reflective answers, not simply ones dictated by your group. There are, of course, cases where this kind of approach seems unlikely to help. In some disputes, the different parties' values are so diametrically opposed that the acceptability of a policy doesn't get closer to a consensus even after separately establishing facts using experts. A decision to change a region's official language with an evenly split population would likely be an example of this.

Rob struggled to help people separate facts from values in his research on the effects of drug prohibitions on society. For example, with economists Peter Reuter and Tom Schelling, he produced a paper offering a detailed taxonomy of drug-related harms ranging from overdose deaths to traffic accidents to violence among drug dealers.[119] They classified each harm as primarily due to the effects of the drug or primarily due to drug prohibition. For example, drugs are a source of transmission of HIV because of drug prohibition (specifically, the lack of access to clean needles), whereas traffic accidents are due to the psychoactive effects of the drug itself. The authors believed, somewhat naively, that this taxonomy would help legalizers and prohibitionists have more productive and dispassionate debates about the relative merits of each position. They expected more pushback from prohibitionists, but most of the complaints came from legalizers. (It may be that prohibitionists never actually read the paper!) Legalizers complained, persuasively, that framing questions about legalization of drug use in terms of harms completely ignored the possible *benefits* of drug use (pleasure, fun, personal exploration, pain relief, etc.)—surely a crucial part of any cost-benefit analysis.

This example may just illustrate a blind spot on Rob's part, but we think it also illustrates the importance of clarity about facts and values in any public debate. As we've just seen, in making big policy decisions, experts are not necessarily authoritative on which factors *matter*, and by how much. Here everyone can contribute.

It certainly isn't always easy to find the key listing of relevant factors that this approach to decision-making requires. But still, the Denver bullet study is instructive. Even when passions on a subject run high, it is possible to take a reflective approach to policymaking that renders our reasoning explicit, and therefore capable of review and discussion.

VALUES DISTORTING FACTUAL CLAIMS AND METHODOLOGIES

Even when the primary factors in a policy decision are clear, there is another concern that can make it difficult to adopt the facts/values approach. During the latter part of the twentieth century, we became increasingly aware of the ways in which our cultures, our goals, and the unequal relationships in our societies can shape our actions and beliefs—and, as we discussed in chapter 13, scientists are not immune to such influences. So while facts are fundamental elements of the shared, objective reality that our efficaciousness depends on, *claims* about facts together with their methodological underpinnings (what we'll call fact stories) can be shaped and distorted by these societal forces. For some, this makes it difficult to imagine separating facts from values in deliberation on policy decisions.

This is not a trivial concern. While we are less worried about the potential for values to distort fact stories for a physics finding like *force equals mass times acceleration,* the situation is less clear for many social science findings. A good example of such a social science fact story where we might worry was highlighted by Alice Eagly and Linda Carli in their 1981 analysis of pooled data from 148 published experiments—on the question of sex differences in conformity to other people's views. Overall, the analysis showed that women were more influenceable than men. (Most of the studies were from the 1970s or earlier and should not be generalized to the present day.) But when they broke the studies down by the gender of the first authors of the studies, they found that the gender conformity effect disappeared in studies by women investigators.

Is it a fact that women (at least prior to 1981) were more conformist than men? Perhaps, but if so, why is the fact only documented by male observers? Are male researchers somehow biased toward tasks or measures that show women as more conformist, or are female researchers somehow biased toward tasks or measures that downplay female conformity? We don't know. But the finding suggests that male and female researchers' views about gender might be influencing their claimed results. The scientists don't doubt that there is a factual reality underneath this confusion—and therefore they are very interested in hunting down where the definitions and choice of experimental tests are being shaped by cultural/gender influences—but at the time of this 1981 study, it was clear that there was a problem with the claims and methodologies of these experiments.

In the abstract, this problem of values distorting *fact stories* has at times been argued as undermining the idea of separating facts themselves from values, but we don't agree. As with many of the concepts discussed in previous chapters, there is a practical value in separating the discussion into those aspects that have a factual basis and those that are primarily dependent on values, goals, and emotions. Even if this separation depends on definitions and classifications that may be challenged, there is much to be gained by working toward this aspiration, since it forces the parties to separately consider the ways in which reality constrains a decision.

VALUES AND CONFLICT

What values do people have? Is it possible to give a systemic description of what matters to people, whoever they are, in a way that we might bring to bear in practical decision-making? One approach is just to *ask* people what they value, and how much. Of course, what people say is one thing, and which values they actually bring to bear in practice is another. But still, asking people about their values isn't beside the point.

Do values always conflict? No, and when they do, it is a matter of degree. In the social sciences, the dominant framework for mapping out value conflicts was developed by Shalom Schwartz.[120] He surveyed people from twenty nations, asking them to rate the importance of fifty-six different values "as a guiding principle in my life." Statistical modeling showed that the core values of power, achievement, hedonism, stimulation, self-direction, universalism, benevolence, conformity/tradition, and security can be either closely allied (people who prioritize one will tend to prioritize the other), diametrically opposed (people who prioritize one will tend *not* to prioritize the other), or unrelated to one another. That is, for most people, some of these values (hedonism and tradition) will be very difficult to reconcile with each other if a decision affects both of them, since they tend to push in opposite directions. Other pairs of values (such as power and achievement) can both be promoted at the same time—they won't push the policy decisions in opposite directions. This characterization of the variety of values has spawned a large international research literature analyzing real-world intergroup conflicts in domains ranging from immigration to climate change to ethnic conflicts in many regions of the world.

Philip Tetlock's value pluralism model[121] argues that value conflicts (for example, the need to make trade-offs between equality and economic efficiency) are psychologically aversive. People have different strategies for coping with these uncomfortable trade-offs. They may engage in denial, simply ignoring one of the conflicting values. If they are accountable to an audience of mixed or unknown views, they may engage in buck-passing (letting someone else make the decision) or procrastination. If they are accountable to a homogenous audience with a known point of view, they may engage in oppositional posturing, telling the audience what they want to hear and demonizing the (mostly nonpresent) opponents.

But there may be productive ways to take some of the sting out of value trade-offs. Claude Steele argues that the ego-defensive and value-expressive functions of attitudes come from a psychological *self-affirmation* system that strives to maintain a positive view of our self as morally good, reasonable, independent, and competent.[122] Being confronted with information that threatens this view triggers opposition, denial, rationalization, and other efforts to reject the information. So far, Steele's ideas sound consistent with what we've just discussed about value trade-offs and attitude functions. But what's particularly interesting is an implication that Steele drew from these ideas, and a procedure for constructively dealing with value conflicts. Steele reasoned that because information that challenges our views is threatening, it is possible to make the individual more resilient and better able to withstand the threat by allowing them an opportunity to affirm their values. In a variety of experiments, Steele and his colleagues have demonstrated that people who encounter new evidence will more readily consider it, and even change their mind, if they first have an opportunity to openly endorse their core values—for example, by completing a values inventory questionnaire—which reduces their need to rigidly defend their values. Researchers have implemented self-affirmation programs in various real-world settings like schools, showing that regular self-affirmations can increase students' willingness to absorb new information in a way that improves their academic performance.[123]

SHARED VALUES BEYOND BADGING

These elements and perspectives on our relationship to our values and their origins lead us to ask a somewhat bigger question: How is it that we sometimes do seem to make collective progress on a shared understanding of values, even though we can start from very different cultural assumptions and different badged identities? For example, over the centuries people have convinced one another of key consensus values. Who today would debate the evilness of slavery, rape, or intentional humiliation of vulnerable people? Apparently, there are routes to shared values.

So how do we reason about matters of value? We talked in earlier chapters about how the metaphor of the raft works in scientific reasoning. We can see another way the metaphor of the raft is useful here, in describing our thinking about values. In fact, children begin reasoning about values from a very early age, often in dialogue with a parent:

Child: "Ting-An lent me her pen, but I don't think I'll give it back to her. If I give it back to her, then I won't have a pen anymore. So why should I do that?"
Parent: "So you don't think people should give back things they've borrowed?"
Child: "Well, maybe they should usually, but this is different."
Parent: "Why is it different?"
Child: "Because Ting-An's younger than me."
Parent: "But if people don't have to give back things they borrowed from younger people, then James shouldn't have to give you back the spinner he borrowed from you."

And so on. The reasoning here moves back and forth from (1) judgments about what's okay in particular cases (giving Ting-An back her pen, James giving you back your spinner) to (2) judgments about general rules governing behavior (people should give back things they borrowed). We make decisions about what's okay in a particular case, but we expect to have to confirm these by general principles that are illustrated by the particular case. If the general principle seems okay, we might leave it there. But if the general principle doesn't seem okay—you find yourself, for example, arguing that no one has to give back things they borrowed from someone younger than them—you have to either withdraw your judgment about the particular case (it wasn't okay to keep the pen after all) or reframe the general principle until you find one that does seem right. We bounce back and forth between our judgments about particular cases and our judgments about general rules, and we act when we have reached a settled state, where we're comfortable both with our actions in particular cases, and with the general rules that our actions illustrate. This is the method of "reflective equilibrium."[124]

A similar process happens among competent adults, except that the teaching goes in both directions (and adults at least know that they *should* take other people's interests into account). We learn from one another, and even learn across cultures. For example, if someone didn't understand what's wrong with slavery or humiliation, one way to educate them would be to show them a particular, graphic example, and have them agree that it is not okay. Then we would work backward to find out what general principles are in play here—that human beings are not entitled to own or brutalize or do violence to other human beings; that might does not make right, etc.—and having formulated the principles, we can review whether they seem okay as they stand, or we can look at more cases and the verdicts on them delivered by our principles.

The raft metaphor we used earlier is helpful here, too. We don't have to start from ground zero in our thinking about values and justify everything from scratch. We start with the raft of value judgments provided by our society and use our common sense to think about what's okay to do and what isn't. And while remaining afloat on that assemblage of value judgments, we can take some out, one by one, and inspect them to see if they are really right. If it's a general principle, we can review it to see if it generates judgments about what's okay in particular cases that seem right. If it's a judgment about a particular case, we can ask what general principle is being used there, to see if that general principle seems sound. So there is here a role for iterative deliberation concerning values, just as we use iterative deliberation to adjudicate among different sources of evidence and expertise when we are determining facts.

Practically speaking, these discussions and recent findings suggest ways to improve the values part of a deliberation:

- First of all, expect many iterations of "reflective equilibrium," between you and the people you're talking with, as your discussions move back and forth between specific examples, where the participants have emotional reactions, and general principles that might explain these reactions.
- Provide respectful opportunities for participants to openly discuss values they think are important, early in the process.

- When you and your opponent are arguing on behalf of different values that appear to be in tension with one another, recognize that you may both share each of these values even if you differ in how you prioritize them, and look for creative ways to discuss them.

THE VALUE OF DELIBERATION

The philosopher Peter Strawson once wrote:

> A Scottish judge of the late eighteenth century, asked how he arrived at his decisions, is said to have replied to somewhat the following effect: "I first read all the pleadings and then, after letting them jumble about in my head (with the toddy) for two or three days, I give my judgment." But perhaps he was not a very good judge.[125]

That is not a bad description of what many of us do: when presented with a referendum, we read what we can about a policy, then we let it all jumble about in our heads for a few days, then we vote. This chapter's facts-versus-values discussions should help make it painfully obvious how much this approach very often misses: we aren't explicit about the elements of a policy decision, we really don't know how to review it, and we don't know how to argue with someone who disagrees with us about the bottom-line decision.

We began this chapter with a plausibly practical procedure for a group to use to make a policy decision where both facts and values are in play. Perhaps it was an improvement over letting the arguments "jumble about in my head," but the example of the Denver bullet study still doesn't serve as an ideal model for a process that distinguishes facts from values and incorporates both. Its algebraic, algorithmic procedure doesn't include some of the principled elements that we would like to see in a deliberation both for the facts side and for the values side. For example, rather than just averaging the experts' factual numbers or the weights of the citizens' values, there should be room for iterative deliberation on these sources of input, perhaps first using Steele's self-affirmation strategy. Such dynamic deliberations are the topic of our next chapter.

CHAPTER 17

The Deliberation Challenge

To kick off this penultimate chapter, we would like to offer a startling proposition and a sober-minded challenge. The proposition first: *We may be the first generations in human history that could reasonably aim to build a lasting world in which every person can thrive.* This proposition is certainly debatable, maybe even unlikely, but just the possibility that it is true should wake us up. What do we mean by it? First, as we discussed in chapter 10, it is only our generations—the people alive today—that have been able to see the consequences of worldwide efforts to feed the planet: Just since the last four decades of the twentieth century, the fraction of the world's people living in extreme poverty has dropped from over half to less than a tenth, despite the 2.5 times larger population. We have seen world literacy rates climb from under half to 87 percent in that same time frame. Moreover, for the first time in history we have seen population increases slowing down, and in many countries—including the long-time largest in the world—populations have even begun to shrink. So providing for everyone is newly plausible; we don't live in a Malthusian world—one destined to contain more people than resources necessary to sustain them.

The most recent pandemic has been scary, but it has also demonstrated that we can use our rapidly advancing biological knowledge to construct a vaccination on the fly. We have a way to go before we are confident that we can handle each such future threat in time, but we certainly seem to be en route.

On an even bigger, planetary scale, we are the first generations in the first species that is capable of intentionally shaping our global environment. Yes, human industry can cause global warming, but our very capacity to do so also signals that we are now operating at a global scale, no longer simply the passive recipients of a fluctuating planetary environment. Previous peoples have risen and fallen as ice ages (or droughts) have shaped our history, and though—to be very clear—we do not yet know how to safely manage our climate to stabilize these fluctuations, we do for the first time have tools to consider (and invent) that could make a difference the next time glaciers encroach on our populations.

We are also apparently the first species plausibly capable of preventing the next of the mass extinctions that occur every 26 million years or so when a large comet or asteroid has collided with the earth, wiping out most families of species. We have built telescopes that can detect incoming comets and asteroids well before they arrive, and we have practiced sending a spacecraft to nudge such a distant asteroid so that it would miss us.

In sum, we don't yet know how to do the job, but we do know that the generations alive today have the inspiring opportunity to build a thriving world that can outlast all but the microorganisms. We assume that as you consider this startling proposition, you're likely having the same reaction that we have: "Yes, but...! Well, maybe, but...!" It's all very well to use our scientific-optimism-based persistence, but clearly the world isn't the paradise of these dreams. We're not even using our *current* knowledge as well as we could, let alone showing ourselves ready to take on this idyllic goal.

This brings us to describing the sober-minded challenge that we promised at the beginning of this chapter. Arguably, considering this goal from our current perspective, the most important tools missing from our tool chest are techniques for constructive, large-scale collective thinking. When people think well together, we can accomplish amazing, seemingly impossible tasks. When we don't, we rapidly reach logjams—or worse, become destructive. Our collective challenge—possibly *the* grand challenge of our times—is to invent such tools to think together productively, and then to use them. If we can meet this challenge, here at the beginning of the Third Millennium, we have a fighting chance to build the foundation for a thriving planet.

Once we humans have identified a problem, we have been remarkably capable of inventing ways to address it. We also have some starting points to work from if we want to foster productive collective thinking. Just as we recognized that our industry has the ability to affect our global climate (for bad or for good), we should also notice that our industrial-scale information technology can have a national, even global effect on collective thinking.

For example, in the past decade, it has become apparent that our information technologies make it possible to narrowcast messages in ways that create "thought echo chambers," where we only hear messages that reconfirm our viewpoints. This is a great amplifier of confirmation bias, and has clearly led to a partisan polarization that prevents us from making progress on many problems, in many countries around the world. But if we have the ability to generate polarized thought communities, we should also be able to generate more productive collective-thinking environments. We just need to figure out how to use our technology differently.

Throughout this book we have discussed approaches, tricks of the trade, and habits of mind that make us much more effective as individuals, but that can also make us successful collectively. As we focus on the big challenge of developing specific tools for collective thinking that are effective in our digital world, we will again find that they have both a big and a small aspect. Every approach we invent can help us build a thriving planet, but it also can help us build a thriving city, or corporation, or nonprofit—or even a family or friendship group.

DELIBERATION TECHNOLOGY

In order to take up this challenge, it can be extremely helpful to see examples of successful, effective collective thinking. Even if each example by itself doesn't answer our full need, it does show us something we can build on. In the previous chapter, we saw an example of a facts/values weighting technique that could be useful in certain situations. It wasn't clear, however, that the Denver bullet study incorporated enough of the elements necessary for constructive collective thinking to become an important tool in our 3MT toolbox. Let's look at a technique that could play a bigger role. Called Deliberative Polling, it is one of the most inspiring examples we've seen of a transformative tool for productive collective thinking.

The Denver bullet study highlighted the difficulty of bringing facts into discussions in which the people involved have different values and priorities. In such cases, fears, desires, ambitions, and goals tend to drive people's views. Certainly, this is the case in our current political moment, such that people often feel they have no way to talk to one another. Those on different poles of the political spectrum completely disagree, so nearly half the country feels like they've been completely pushed out of the political system by one election. Then another election comes along, and the other half of the country feels pushed out. This can't be a good way to run a country. There's got to be a way to get people to weave together all their different interests, goals, and desires with some factual expertise on whatever issues they seek to resolve, as they make decisions, form plans, and agree on policies. That's what Deliberative Polling is designed to do.

Deliberative Polling was developed in the late 1980s by Jim Fishkin, who was a professor at the University of Texas at the time. He then moved to Stanford, where he now runs the Deliberative Democracy Lab. When Fishkin describes how he came up with the idea of Deliberative Polling, he says it came from thinking about what happens when you take a typical poll. A polling or media outfit—let's say the *New York Times*—asks 1,000 or so randomly chosen members of the American public a question like, "Should America sign a trans-Pacific trade agreement?" Most people probably initially think, "Well, I don't know." But, perhaps surprisingly, the typical person still can come up with an opinion even on topics they don't know anything about, so they will say something like, "Okay... yes" (or "no").[126] The next day the *New York Times* headline says "The American public favors the trade agreement two to one." If you consider how the poll was conducted, you can't help thinking that's a pretty useless conclusion. The people being polled likely don't know much if anything about the topic; they're just telling you what they might think they're supposed to say.

What you really want to know is what a representative sample of United States citizens would think about an issue if they actually knew something about it. Ideally, you'd want to know what they thought after they'd become informed about it and carefully considered the different options and their consequences. Fishkin asked himself, Well, why don't we try that? Why don't we fly a representative sample of United States citizens to a convention center and create a process whereby the participants can become more knowledgeable while they consider the issue, gathering information from both experts and other participating citizens via discussion and deliberation?

Over time, Fishkin and his colleagues developed a process that goes something like this: They randomly select several hundred citizens and bring them together, typically for a three-day event, to deliberate on a specific policy issue like welfare reform. First, they take a pre poll to gauge people's knowledge and opinions about the core issues. For example, in the welfare reform case, they might ask each deliberator questions about their political leanings on the subject, their knowledge about basic economic theory, etc. Once initial opinions are gathered, they divide the participants into small groups of about a dozen people, more or less, each with a trained moderator. Then they start deliberating, based on carefully prepared briefing materials that they have all read (or viewed in video form). The briefing materials aim to present the policy issue, describing the agreed and disputed facts and values, the evidence for these positions, and the pro and con arguments for the policies under consideration—all developed with the advice of experts from all sides of the issue. The deliberation, while open-ended, is moderated so as to maximize productive discussion. The moderator is not allowed to add any content, merely to make sure each deliberator gets a chance to say their piece or to point the group to relevant information in the briefing materials. The moderator also ensures that no voting is conducted during deliberation; quite unlike many jury trials we see in movies, deliberative groups are not allowed to vote on anything during deliberation—except what questions to ask the experts.

Typically, each group reaches a point in its deliberation when its members can't decide on the right choice. So the groups identify all their unanswered questions about what would happen if a particular policy were instituted: how could you prevent unintended consequences, how might it be possible to encourage a certain outcome, etc. The groups collect these questions and present them to a panel of experts covering the full gamut of views on the topic.

The panelists represent different positions, so of course they often disagree. But they all have some expertise on the issue. The experts are not allowed to lecture the deliberators; they're not supposed to try to change their opinions. They're just there to answer the questions. The experts are grilled by the groups, and then the small-group deliberations resume. The deliberative groups now have more information to work with. Someone might say, "Well, that expert said such and such." And somebody else might respond, "Yes, but that doesn't make any sense if you consider what that other expert pointed out." After this kind of discussion, the deliberative group usually develops a sense for which expert is more knowledgeable about a given topic—and, ideally, which expert is better able to recognize when they don't know an answer and correctly determine when their confidence level on a proposition should be low (that is, in the language of chapter 5, which expert is "well calibrated"). As they continue to deliberate, they come up with more questions. And they go back to the panel of experts. They go through this iterative process a number of times over the course of the long weekend. The goal is not to come to consensus. It's not like a jury, where everyone has to agree. However, by the end of the event, people have often changed their minds. Fishkin and his colleagues can track and quantify what happens to opinions because they poll participants both before and after the event. They've found that participants do not change their minds based on who is the most charismatic expert or who is the most well-spoken or has the highest status in the deliberation group. They change their minds based on the new information they get about the problem.

After these deliberations it's common for people to say they came in with a particular view and then realized they were just wrong. Someone in their group might have talked about a relative who was in a particular situation, and that would personalize the issue enough to make them realize there were other perspectives to consider. Or, after listening to the experts, they might realize that they were not taking into account a certain fact that was an important part of the story.

SIDELINING APATHY AND "BADGING"

Deliberative Polling may also be an encouraging answer to the apathy and disengagement that get in the way of democracy. In the Deliberative Polling events that Fishkin and his colleagues have conducted, people tend to take their participation seriously. Fishkin and his colleagues have at times achieved remarkably high participation rates among the people they've randomly selected to join, and more than 95 percent of those who start the deliberations complete them. What's more, they find that the people who participate tend to increase their exposure to news media before the event. They have stories of people, for example, who'd never read a newspaper and start reading three newspapers a day. People take the invitation to participate as a real duty to do a good job.

The phenomenon of badging that we discussed in chapter 12 appears to fall by the wayside too. During the Deliberative Polling event, people start to identify with the group of people that they're deliberating with and put aside their identity as, for example, a die-hard liberal or an economic conservative. These badges don't become the dominant factor when people deliberate in this way.

Hearing Fishkin describe the results of his Deliberative Polling events leaves one feeling much more optimistic about what would happen if you asked a randomly selected "jury" of people drawn from the American public to work together, informed by expertise, on solving a problem, rather than relying on our elected representatives alone. The people we elect to represent us often feel the pressure to appeal to the segment of the population most aligned with them ideologically, rather than representing the interests of everyone in their district or state (let alone the country).

Could we come up with ways to reproduce the effects of a Deliberative Polling event on a much larger scale? For example, might it work with a great number of small videoconferences online? Or perhaps a consortium of philanthropic foundations could run a few national in-person Deliberative Polls each year and then spend a substantial amount of money on advertising how and why people changed their minds on the issues discussed, to stimulate national thought on the topic? Such large-scale Deliberative Polling efforts might help people become better at generating an unbiased understanding of the facts, taking them into account, and finding compromises—not to mention listening to and understanding the views of others. This could translate into better-informed voters and legislators.

SCENARIO PLANNING

Deliberative Polling is inspiring in its ability to bring together expertise with the legitimacy—the right to be heeded—of a random sample of the population that's large enough to fairly represent the values and emotions of the stakeholders. But what about planning decisions where much more is unknown, where expertise only goes so far, where the unforeseeable future state of the world could change what we would decide? The technique of "scenario planning" was invented for just such situations.

Scenario planning was initially developed in the 1960s by Herman Kahn at the RAND Corporation and later the Hudson Institute. (Kahn was a major inspiration for Peter Sellers's title character in Stanley Kubrick's 1964 satire *Dr. Strangelove.*) Other early developers were the Stanford Research Institute and the Royal Dutch Shell Group, especially during the tenure of Peter Schwartz as head of their planning division. In *The Art of the Long View,* Schwartz describes how scenario planning was further developed for business applications, and ended up being successfully used for dramatically varied purposes, from the "sublime"—planning the rebuilding of a society after apartheid ended in South Africa—to the "ridiculous"—creating a future for the science fiction movie *Minority Report.*

The basic concept of scenario planning is to develop a range of possible futures so we can test how robust our decisions are in these contexts.[127] The participants in a scenario planning exercise begin by identifying a pending important decision that needs to be made. (It is possible to do scenario planning without any pending decisions in mind, but a focus on concrete decisions makes the exercise less abstract and more useful.) For example, a business may need to decide what workforce changes it will need in the upcoming decades, or, more personally, a college student may need to decide what field to pursue and what trainings to complete.

Participants then identify "key forces" in the immediate, local environment, as well as "driving forces" in the larger macroenvironment, relevant to the pending decision. Because there is no limit to the number of potential forces in the local and macroenvironments, participants are encouraged to narrow the list to those factors that seem most likely to determine the ultimate success or failure of the decision. For the workforce/training decisions, driving forces might include:

- High-quality education (free and universal vs. expensive and restricted to elites?)
- The economy (growth vs. stagnation or depression?)
- Distribution of wealth and power (highly concentrated vs. more evenly distributed?)
- Age distribution (will it continue to shift toward an older population?)
- AI and robotics (how will they play a more dominant role than today?)
- Future pandemics (and our ability to manage them)
- Possible culture shifts in work-life balance
- Degree of globalization (interdependent/low conflict vs. isolationist/high conflict?)
- Energy costs (becoming negligible versus becoming more expensive?)

Next, participants rank each force with respect to its importance and its uncertainty. Driving forces that are very uncertain are more useful to consider, since forces that are either "impossible" or "predetermined" will be common to all the scenarios you will generate (in the next step). For example, there is probably more uncertainty about future energy costs than about the shift in demographics toward a more aged population, so this would make the energy dimension more useful to explore than that of the demographic shift, which will occur in all of your future scenarios.

Although there may be relevant data pertaining to the different driving forces, it would be a mistake to allow the availability of data to influence what gets ranked and how it gets ranked, because the goal here is very different from the goal when we are developing our first-order and second-order causal factors. There, we want to know what the current reality is, so data can be very helpful in ranking the factors, whereas here, we want to consider a wide range of possibilities for what the future might hold, including unlikely ones. It is the fact that we are considering a wide range that helps in our planning, even if the particular range we explore doesn't cover every possible future.

Participants then select and develop a small number of future scenarios. As is the case for people writing fiction or developing screenplays, there is no limit to the number or complexity of the scenarios that participants can generate, but scenario planning is most effective when they develop a limited number, typically four. These four scenarios are usually based on a matrix of the top two driving forces, each broken into the directions that the driving force could go in the future (e.g., "Wealth is evenly distributed" versus "Wealth is highly concentrated"). Thus, for two key driving forces and two levels of each force, there are 2 x 2, or 4, cells in the matrix. Each of these cells then represents one part of the range of possibilities that seem particularly consequential for the success or failure of the decision. Participants are encouraged to give each scenario a pithy and evocative name that is memorable enough to play a role in future deliberations about the decision. The scenarios are turned into brief narratives, much like imaginary news stories after an event has taken place. Each narrative should explain how the scenario, even if unlikely, could plausibly play out.

For example, for our business work force and student training decisions, imagine that we chose "Wealth concentration" as one of our key driving forces and "Much more effective AI and robotics" as the other. Then the matrix of scenarios might end up looking like this:

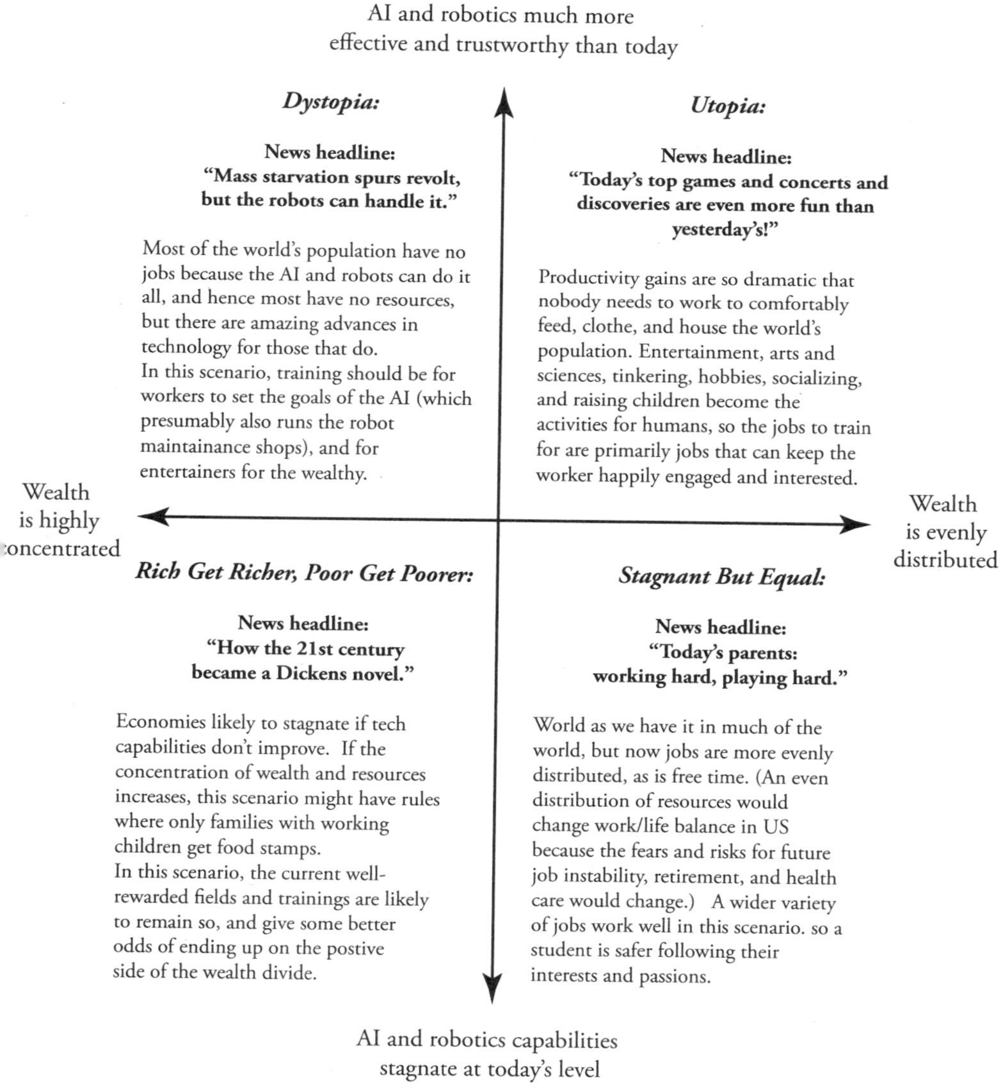

AI and robotics much more effective and trustworthy than today

Dystopia:

News headline:
"Mass starvation spurs revolt, but the robots can handle it."

Most of the world's population have no jobs because the AI and robots can do it all, and hence most have no resources, but there are amazing advances in technology for those that do.
In this scenario, training should be for workers to set the goals of the AI (which presumably also runs the robot maintainance shops), and for entertainers for the wealthy.

Utopia:

News headline:
"Today's top games and concerts and discoveries are even more fun than yesterday's!"

Productivity gains are so dramatic that nobody needs to work to comfortably feed, clothe, and house the world's population. Entertainment, arts and sciences, tinkering, hobbies, socializing, and raising children become the activities for humans, so the jobs to train for are primarily jobs that can keep the worker happily engaged and interested.

Wealth is highly concentrated

Wealth is evenly distributed

Rich Get Richer, Poor Get Poorer:

News headline:
"How the 21st century became a Dickens novel."

Economies likely to stagnate if tech capabilities don't improve. If the concentration of wealth and resources increases, this scenario might have rules where only families with working children get food stamps.
In this scenario, the current well-rewarded fields and trainings are likely to remain so, and give some better odds of ending up on the postive side of the wealth divide.

Stagnant But Equal:

News headline:
"Today's parents: working hard, playing hard."

World as we have it in much of the world, but now jobs are more evenly distributed, as is free time. (An even distribution of resources would change work/life balance in US because the fears and risks for future job instability, retirement, and health care would change.) A wider variety of jobs work well in this scenario. so a student is safer following their interests and passions.

AI and robotics capabilities stagnate at today's level

At the end of the process, participants consider whether each of the decision options they are considering will be (as Schwartz puts it) "robust across all scenarios." The most robust option may be least risky, but it is not necessarily the optimal choice; a less robust option might be preferable because it would produce the best possible outcome under one scenario, and we are willing to take a little more risk that one of the other scenarios ends up being the world of the future. (Someone who is risk-averse might choose to get trained in AI goal-setting if that looked like a robust career choice across all scenarios, but someone with a higher tolerance for risk might choose a creative job path, like architecture, if they thought that would be optimal in all but, for example, the bottom-left scenario.) But the exercise of identifying the robustness of each choice is valuable for mitigating the kind of wishful and distorted thinking we described in our earlier discussion of "groupthink."

Finally, to help make the scenarios concrete, and in order to monitor the outcome of the decision, the participants use their list of key forces to develop a set of useful indicators or signposts for identifying whether any of the scenarios appears to be coming true. If newspaper headlines appear in the next few year announcing breakthroughs in AI that make it more trustworthy and effective, and if "universal basic income" becomes part of the political platforms of all parties, then one might start to infer that the bottom-left scenario is not as likely.

By design, scenario planning also serves as a good antidote to confirmation bias; it has a lot in common with the "consider the opposite" method of debiasing we mentioned in chapter 12.

FORECASTING TECHNOLOGIES

Scenario planning is a useful tool for stimulating careful thinking and planning about the future, a frequent goal of collective thinking and decision-making. But it is not intended to forecast the most likely scenario that we will find ourselves living in. In order to do that, several collective-forecasting methods have been developed. The Delphi Method is a prominent early example developed at RAND during the Cold War era. There are many variations, but the basic idea builds on Francis

Galton's early demonstration of the wisdom of crowds—the idea that an aggregation of individual judgments tends to be more accurate than most (or all) individual judgments. Experts in relevant topics are asked to offer their best quantitative rating or prediction for the focal question. They do so anonymously to promote candor and to discourage either timidity or grandstanding. Galton's method was simply to solicit ratings or prediction once. But Delphi usually involves multiple iterations, in which the judgments are then circulated back to the participants, who can decide whether to revise their judgments in the light of the other opinions. In one version of the method, this process simply repeats for several rounds, and the final product is the average of the final set, with no discussion ever taking place. In another variation, experts deliberate to resolve any remaining disagreements and seek consensus. As we've discussed, allowing people to adjust to others' opinions can work well if experts have a good reason for changing their judgment—and share a conceptual scheme that allows them to identify the best of the submitted judgments (a truth wins process). But it doesn't work as well as simply averaging the expert views if an expert with a very accurate judgment shifts to conform to the rest of the crowd (as can happen in a majority wins process). A more recent twist on collective-forecasting methodology is the modern *prediction market.* Gamblers have bet on the outcomes of popular elections at least since the early twentieth century (and probably far longer), but the modern institutionalized practice was largely developed by business school professors at the University of Iowa in the late 1980s before being commercialized by for-profit firms such as Intrade and PredictIt. In ordinary "wisdom of crowds" demonstrations, participants have no incentive for accuracy other than the desire to do well. In a prediction market, participants put their money where their mouth is by purchasing a contract for a particular outcome (e.g., that a given candidate will win the next election), analogous to commodities trading for pork bellies or other agricultural products. Excitement about prediction markets grew over the early 2000s after these markets outperformed professional polling firms in predicting the 1998 US national election. Various studies showed that prediction markets often outperformed the average of the polls (a method which itself already improves poll accuracy by canceling out some error in individual poll results). Prediction markets quickly came to be seen as a perfect synergy of the wisdom of crowds phenomenon and economists' "efficient market hypothesis," which roughly states that the markets tend toward optimal performance by taking all relevant information into account.

Over the last decade or two, prediction markets have lost some of their early promise. They mistakenly predicted that weapons of mass destruction would be found in the first Gulf War, and they failed to predict Brexit or Donald Trump's victory over Hillary Clinton. More recently, they overpredicted (like everyone else) the Republican Party's performance in the 2022 US midterm elections. Nevertheless, the two conceptual foundations of prediction markets—the error-canceling wisdom of crowds phenomenon and the efficiency of markets—are "long-run" benefits, even if not necessarily realized in every individual case. It seems a safe bet that prediction markets are here to stay and demonstrate useful approaches to collective thinking.

Our final forecasting example is the remarkable Good Judgment Project (GJP), developed by Phil Tetlock and Barb Mellers at the Wharton School. Over the course of their careers, Tetlock and Mellers made major contributions to our understanding of the strengths and foibles of human judgment. In 2005, Tetlock stirred up controversy with a book claiming that, according to carefully conducted tests, professional political experts were no better at making specific predictions than we'd expect from a "dart-throwing chimp." (Tetlock was not arguing that these experts are useless; rather, that the value they add comes from helping us to understand the world, and that the complex webs of multiple causation that affect that world make them not so good at predicting which of two possible outcomes will actually occur.) Surprisingly, just a decade later, Tetlock's 2015 book *Superforecasting* (with Dan Gardner) tells a much more encouraging story about what political forecasting can achieve using methods the GJP developed.

The GJP emerged out of a series of political forecasting tournaments sponsored by the Intelligence Advanced Research Projects Activity (IARPA), a research branch of the US intelligence community. With their eyes wide open about the pitfalls of human judgment, Tetlock and Mellers developed a method for building on human judgmental strengths. The bottom line: their method works, spectacularly. The GJP team won the initial tournament in 2011 as well as every subsequent tournament that was held. The GJP approach outperforms the wisdom of crowds approach (simple aggregation), prediction markets, and other Delphi-like group processes—although it selectively draws on aspects of all three. Startlingly, it outperforms the predictions of professional intelligence analysts who are privy to classified information that GJP lacks.

The entire GJP method protocol is too complex to describe here, but we can give a rough sketch of key features. GJP has an open recruitment method—participants need no special academic or professional credentials. They are incentivized for good performance by a public leaderboard that allows them to compare their performance to their peers in real time. They make quantified predictions (the probability that a particular discrete event will occur by a certain date), and they provide explanations for their predictions. They can update their predictions after receiving new information or reviewing their peers' predictions and explanations. The leaderboard uses a scoring method that rewards them for showing both good accuracy and good calibration. The GJP skims off the consistent top performers (not just those who "luck out" on any particular prediction) to form teams of "superforecasters." Importantly, these superforecasters are often ordinary people without special credentials.

Much remains to be learned about the seemingly magical success of the GJP, but according to Tetlock and Mellers, here are key takeaways so far:

> We discovered four such drivers [of superforecasters' better accuracy]: (a) recruitment and retention of better forecasters (accounting for roughly 10 percent of the advantage of GJP forecasters over those in other research programs); (b) cognitive-debiasing training (accounting for about a 10 percent advantage of the training condition over the no-training condition); (c) more engaging work environments, in the form of collaborative teamwork and prediction markets (accounting for a roughly 10 percent boost relative to

> forecasters working alone); and (d) better statistical methods of distilling the wisdom of the crowd—and winnowing out the madness (. . . which contributed an additional 35 percent boost above unweighted averaging of forecasts).[128]

What distinguished superforecasters from other forecasters? As noted, they didn't need fancy credentials, although they tended to have a lot of political knowledge and to score well on mental ability tests. But they also had open-minded cognitive styles, with a willingness to acknowledge the limits of their knowledge, the weaknesses of their arguments, and the possibility that they might need to revise their beliefs as they learned more. This was reflected in their calibration scores, which showed them to be less overconfident than the regular forecasters.

This description of the superforecasters and their processes should sound familiar to you by now. The superforecasters appear to have been using many of the elements of Third Millennium Thinking that we have been discussing—and quite successfully. Effective collective thinking in action!

ONLINE THINKING TECHNOLOGY

So far, we have looked at collective-thinking examples that have their roots in interpersonal, face-to-face conversations, and we hope to invent ways to scale them to the larger online world. But there are also collective-thinking approaches that have been developed specifically for the online environment. Generally, the goal is to come up with algorithmic systems that would specifically avoid re-creating the kinds of online conditions that work against productive online discourse—the echo chambers of thought, the deliberate reinforcement of "badging," the increasingly angry comment threads.

Several years ago, Saul was talking to Berkeley engineering professor Ken Goldberg about an online deliberation system developed by Ken's research group that helped participants recognize how their positions on various issues compared to those of others using the same deliberation system. Saul asked whether it was possible to test out an alternative algorithm for such a system that would encourage people to understand the opposing position and not just identify with others who share their own position. Could people win points for expressing well the opposing point of view?

In the end, they came up with a demonstration system, called DebateCAFE, which asked people to enter the best arguments they could think of on both sides of a particular issue, and also to score other people's arguments. The unusual element of the system was how you won a place on the leaderboard: your position was based on the worse of your own two scores (the two averages of the ratings other users gave your arguments for each side of an issue), so you had a strong incentive to make both arguments strong.

Another approach to group collaborative thinking is to decompose the elements of a discussion in, say, a news article into small "testable" parts. Does this sentence show evidence of clear probabilistic thinking, or does it fall into one of the probabilistic-thinking mental traps we discussed earlier, like the "look elsewhere effect" or "p-hacking"? Does this other paragraph show an ability to "consider the opposite" or is it a classical confirmation-bias selection of evidence? If each separate question can be asked of a representative group from the whole population of citizens, we could check that there is a consensus on these small points, shared by people from all parts of the political spectrum. Effectively, they would be saying the equivalent of "I like [or dislike] the results of this argument, but at least it doesn't have a problem with p-hacking." Then, if we can put together all the evaluations of this kind for one argument, website, or newspaper article, we can get a sense of its strength and trustworthiness.

It sounds like it would be a lot of work to analyze any one position statement, website, or newspaper article using this approach, but "many hands make light work"; and there are in fact citizen-scientist websites that are set up to algorithmically share such projects among hundreds and thousands of interested citizens. Usually, such websites tackle big science problems, like identifying certain characteristics of hundreds of thousands of galaxies seen in images from the Hubble Space Telescope. Saul had experience with these when he and his collaborators used them in a supernova search project, so he thought there should be at least one such citizen-scientist website to try this approach for newspaper articles. This led to a collaboration between the Berkeley Institute for Data Science where Saul was director and Goodly Labs, a nonprofit run by Nicholas Brigham Adams, who has been leading the effort to build the Public Editor website to try out this concept.[129]

A COMPLICATED AND INSPIRING CHALLENGE FOR OUR TIMES

Have we solved the sober-minded deliberative challenge that we posed with any of these examples? We aren't there yet. Each of these approaches gets at some but not all of what we need. The Denver bullet study's method applied experts to the facts and citizens to the values, but missed, for example, the authority and impartiality of the truly random representative set of participants that Deliberative Polling offers. It also didn't fully develop the facilitated stages of small-group deliberations that Deliberative Polling emphasizes for these representative participants. Meanwhile, Deliberative Polling doesn't currently guide the participants to separate their factual concerns from their values concerns—and neither Deliberative Polling nor the Denver bullet study method further breaks down the arguments into more bite-size testable parts to foster agreement on these components, as Public Editor does. If we care about citizens understanding the points of view that they disagree with so they can truly take advantage of the "marketplace of ideas"—one of the longstanding rationales for freedom of expression in US jurisprudence—then our list of methods only offers DebateCAFE.

Only two of our methods address collective thinking when the most important factors depend on what the future holds: scenario planning and GJP superforecasters. And here we find that only scenario planning emphasizes the ability to plan for a wide range of outcomes, including unlikely but possible ones—certainly an important goal in a world where such surprising events have often disrupted our plans. But if we need to make a bet on a specific event occurring in order to make a decision, then we should go with the superforecasters.

There are other goals that we care about for these collective-thinking methods, and they are often in tension with one another. For example, ideally the method would engage every member of the concerned population, not just a small group, since we would like everybody affected by a decision to understand the grounds on which it was made and to feel that it was fairly reached, even if they don't agree with it. This goal is likely to be difficult to achieve with any method that requires real face-to-face, in-person interaction among participants to build the trust and mutual understanding so often necessary in a fair and reasoned decision. Moreover, we live in a world where it is becoming increasingly difficult to be sure that you are seeing authentic information, not made-up material that is generated by AI and/or by people who don't care about misleading you, whether for financial, political, or other reasons. This concern about authenticity may especially call for more in-person and face-to-face methodology in our collective thinking, adding to the difficulties of scaling to a larger population.

This all sounds hard! But this is the sort of problem where a substantial, iterative, scientific approach can make a big difference. The lesson of the half dozen examples that we have described is that each helps facilitate productive collective thinking—and there are many more iterations, variants, and inventions to try. We don't have to get it perfect; we just have to do better (and better). And a useful method doesn't have to meet all our goals; we can design methods for specific needs. Finally, although we have become all too aware of the dangerous potential of the internet to polarize and confuse our public conversation, we have not yet taken advantage of opportunities for better collective thinking that it might present.

We hope to leave you here with a sense of optimism: There is work to be done, but it is not beyond our human capacity. And the good news is that this work has huge payoffs. Almost every concern of our day needs better collective thinking. We opened this chapter with the "startling proposition" that we may be the first generations in human history that could reasonably aim to build a lasting world in which every person can thrive. But certainly there are enough immediate challenges and goals on our way to Utopia to motivate us to dramatically upgrade our collective thinking.

CHAPTER 18

Rebooting Trust for a New Millennium

We began the last chapter with a grandiose vision of a secure, flourishing world that requires new tools for collaborative thinking. But let's finish the book on a somewhat more personal note: Why do each of us individually care about—need—the tools of Third Millennium Thinking? And why have we authors chosen this particular set of ideas to act as a "starter set" for such 3MT?

In answer to the first question—Why should we care?—there are a carrot and a stick. The carrot: There is real pleasure in taking on the world with these tools, and seeing how they help us become more effective. The stick: We have no choice. The world has moved on from the days when small groups of professionals (major news organizations, medical associations, scientific academies, etc.), networked by messengers, mail, and telephone, digested the world's scarce data to give the rest of us the answers we need for for our daily decision-making. Perhaps the signature difference characterizing the Third Millennium is that now we are all "in the game," all networked, like it or not. With all of our direct access to vast universes of data, we are now forced to figure out what facts to base our decisions on, when to do research for ourselves and when to look for experts, what experts are trustworthy (and on which particular topics), and when we might need wise guidance on integrating values.

Clearly, this Third Millennium development is not all bad—or even mostly bad. A world of autonomous thinkers certainly seems better than a world of sheep. We have to take on this challenge, however, just at a time when our national and international conversations have been fragmented and riven by the echo chambers of narrowcast news and social media; when the dramatically effective viral transmission of misinformation and disinformation can cloud our vision; and when the technologies of artificial intelligence threaten to present us with false representations of reality that are more difficult than ever to distinguish.

So why, at this juncture in history, do we choose to offer the particular set of tools we have discussed in this book? The answer, in part, is that these concepts can be thought of as representing the latest version of the scientific practices that have allowed us to transcend so many previous crises of understanding. There are other concepts (and there will be more) that do this, but the ones we have discussed are good enough to get us going, facing head-on our need to track a reality out there that will go on doing its thing whether we like it or not—a reality that will either thwart us or work for us, depending on how well we understand it.

We have to understand the ways in which we fool ourselves about this reality when we don't understand our shaky (noisy) evidence. We have to build correctly on the probabilistic clues that we are given. And we thrive when we know how to benefit from those who disagree with us enough to give us glimpses of where we are stuck in an incorrect viewpoint. Moreover, we can make even greater steps forward when we can choose experts who demonstrate *their* understanding of all of these key concepts; those who, for example, themselves seek out contrary views.

There are many such ideas in this book, but in the table below, we've highlighted some of the most important ones, dividing them roughly into those that each individual can adopt, and those needed when working with others.

Looking at this full set of tools is a bit daunting. (We authors still all too often forget to apply these tools when we need them!) But we should take some encouragement from one of the positive aspects of the changing culture we live in: When we use these tools and adopt these approaches, we have the wind at our back; the larger world around

Habits of Mind	*Habits of Community*
Build better instruments.	State your confidence level.
Try to separate facts from values.	Be skeptical about possible solutions, but use the can-do persistent culture of science to make solutions possible.
Think probabilistically, not in true/false binaries.	Keep each other honest.
Don't be fooled by patterns in random noise.	Agree on a desired balance of risks (false positives and false negatives).
Distinguish noise and bias.	Adopt effective procedures for deliberation.
Be wary of mental shortcuts.	Continually update this table as we discover new ways that we fool ourselves and new habits of mind and community to avoid doing so.
Avoid confirmation bias.	

us—and the scientific thinking embedded in it—is also moving in this direction in some interesting ways. You are likely to have noticed that the tools and concepts we've discussed are being put into action in the world around you.

It's probably easiest to describe this cultural change that we are experiencing—that is shifting how scientific thinking works—by comparing it to previous cultural periods, as we do in the table on the next page. We have organized it around three eras, but we caution that these don't correspond to three millennia. Column 1 describes some of the enormous intellectual gains of the scientific revolution and its aftermath, mostly during the previous millennium. Column 2 describes sources of disenchantment and backlash, building at the end of the twentieth century and seemingly reaching a head in these early years of the new millennium. Column 3 sketches the evolving cultural period that we are now experiencing—and our 3MT vision for healing the rift between science and society.

The table is a thumbnail sketch; we don't claim it is completely original to us, and we certainly don't claim it can stand alone as an authoritative work of intellectual history. Everything in the first two columns has been discussed and dissected by many different scholars in many different disciplines. But we do think the third column describes emerging patterns that have not been fully or widely appreciated yet.

Our new millennium, though still in its infancy, feels different from the one that came before. Although all of its changes can sometimes leave us feeling like we are running on fumes, with less of a shared sense of purpose and direction, these changes are also helping us to address our individual and collective problems, and to begin to build our next steps.

To unpack how we got here, let's look in a little more detail at what was going on during the transitions described in the table. The second half of the last millennium was a remarkable period of human accomplishment. The European Renaissance of the fifteenth and sixteenth centuries saw a new flourishing in the arts and philosophy, and—most relevant for this book—a revolution in science launched by the insights and methods developed by Galileo, Kepler, and Bacon. The seventeenth and eighteenth centuries witnessed a remarkable blossoming of ideas about rationality, empiricism, and moral and political theory that we now call the Enlightenment, with major conceptual advances by Newton and Leibniz, and the emergence of an organized scientific

Scientific Success (pre-20th century)	*Disenchantment and Backlash (transitional pains)*	*The 3MT Renewal*
Instrumentation Experimentation Replication Computation Scientific societies and peer-review journals Single- and double-blind controls Scientific skepticism (claims need hard evidence) Scientific optimism (with can-do persistence, we can discover what is real, and problems can be solved) Scientific realism	Institutional "cloistering" of scientists Homogeneity of experts: • Race, ethnicity, gender • Class • Geographic location • Political views Economic conflicts of interest Expert overconfidence and overclaiming "Build it because we can" and the rise of technological risks: • Nuclear weapons • Climate change • Biohazards • Opioids • AI, nanotech, etc. • Automated warfare • Viral social media • Automated stock exchange	Shift from "factual thinking" to "probabilistic thinking" Shift from "reductionism is all" to a multilevel, nuanced view that includes emergent phenomena Shift from "masterstroke solutions" (progress by great leaps) to "iterative solutions" (progress by careful steps) and an "experimenting society" Shift from "technocratic decision-making" (experts and leaders decide) to "deliberative decision-making" (collective consultation and consensus-seeking) Shift from zero-sum-game trade-offs to more ambitious, more can-do enlarge-the-pie win-win solutions Interdisciplinary teamwork New collective tools: Open science (preregistration, data sharing), blinded analysis, multilab collaborations and verification, citizen science and fact-checking, Deliberative Polling, scenario planning, prediction markets, superforecasting, online platforms for dialogue and debate

community with societies and journals and peer review. The nineteenth and twentieth centuries brought a radical rethinking of fundamental assumptions about time, space, and life by Einstein and Darwin, paralleled by exponential growth in technology due to inventors like Watt, Bell, and Edison. By the close of the millennium, we saw digital technologies radically reshape our economy and our culture.

But as we noted at the outset of this book, by the end of the twentieth century, this bullish sense of progress had become passé. Once-radical advances seemed less astonishing and more routine, and many saw utopian views of human advancement as naive or hopelessly rooted in the former, outmoded and discredited colonial systems of power. Worse still, recent years have seen assaults on scientific optimism far more profound and withering than even the most scathing academic critiques. Faith in two seemingly bedrock assumptions—that assertions need a warrant in evidence, and that scientific investigations are our most powerful method for providing that warrant—seems to be eroding more rapidly than many thought possible.[130] (Even in recent years, scientists and doctors have been ranked among the most trusted professionals in US and international polls, but there has been some partisan-specific slippage over time.)

It is not simply a matter of whether some people are ignorant of science. Even the giddiest about science—and you've surely noticed by now that we count ourselves as giddy science fans—have to acknowledge that while our scientific advances have enabled technological advances, our efficaciousness in modifying the world has come at a price. As the magnitude of our interventions increases, the desired effects get larger, but so do the unintended side effects. Better painkillers come with more addiction. Faster transportation comes with congestion and pollution. More rapid social communication amplifies information but also disinformation.

As members of a society, we don't just want science to increase our knowledge; we want science to solve problems. But perhaps the notion of "solving" problems is confused. The term "solutions" implies a finality that might be illusory. We may wish to fix something once and for all so we can turn our attention to other matters, but maybe it doesn't work that way. Maybe we need to view problems as something we manage through continuous adjustment, like tending a garden or tuning a guitar.

There is a wonderful essay by Donald Campbell titled "The Experimenting Society."[131] Campbell's vision is of a society that "would vigorously try out possible solutions to recurrent problems and would make hard-headed, multidimensional evaluations of outcomes, and when the evaluation of one reform showed it to have been ineffective or harmful, would move on to try other alternatives." Campbell argued that the experimenting society isn't a static structure but an ongoing process "committed to reality-testing, to self-criticism, to avoiding self-deception."

So while the left-hand column of our table of concepts from the book aims to sketch out some of the 3MT "habits of mind" that can foster such a society, the right-hand column, crucially, makes the case that habits of mind aren't enough. They need to be embedded in 3MT "habits of community" in which we both keep one another honest and keep one another from getting discouraged, continuously reminding ourselves that we can do better and that we can get better at doing better.[132]

TRUST, THE FIRST AND LAST FRONTIER

The increasing importance of this "habits of community" list changes the character of Third Millennium Thinking. This is why we didn't write this in the style of a good self-help book (e.g., *Think Like a Scientist and Win! A Guide for Busy Managers, Lawyers, Parents, Doctors, and their Patients*). We do think these thinking tools are in fact practical and useful for your day-to-day life, and we've given many examples. But we see the habits of mind as working together with the habits of community to become part of a bigger picture as well, something big enough to title the book as we did: These ideas anchor the large societal shift in thinking that in the table above we called the 3MT Renewal. And, we hope, they offer our society a route out of the crisis of confidence and trust that is besetting us.

In different times, in different eras, we humans have used different organizing principles as common grounds—trust networks—for discussions and decisions. Sometimes we have used a political or economic structure, for better or worse, such as feudalism set in a monarchy, or capitalism or communism. Sometimes we have built our thinking around a common national culture, with its histories and myths. But as we enter the Third Millennium, we need to work with a much wider array of cultures all at the same time, as we live in much more diverse communities locally and are connected with—and interdependent with—a global society in a way that is more continuous and pervasive in our lives than ever before.

For this next stage of our collective lives, we see the common vocabulary of ideas that we are calling Third Millennium Thinking as a possible shared culture, a next organizing principle, in which to root our Third Millennium discussions and decisions. The self-questioning nature of the 3MT style of collective thinking provides both a need for and a source of trust: our growing understanding of our mental weaknesses—in particular, our ability to fool ourselves—comes along with our growing understanding of how we can tackle this by working with others.

But we should address head-on the elephant in the room (if one talks to elephants): How can an organizing principle for trust be built on a culture that itself depends on trust? Sure, if you feel that you are dealing with people who are basically good-hearted, then the 3MT tools are particularly useful to advance collectively, but of course there are people who will never adopt this culture, who are never willing to be proven wrong. We can't construct a society on the assumption that everybody is operating in good faith. Fortunately, as we will see, we don't have to.

TIT FOR TAT AND SOCIAL OPTIMISM

The challenge of how people of goodwill can succeed and thrive together, in a world that is populated with the usual mixture of well-meaning and selfish characters, is not a new one. The question of how cooperative relationships emerge has preoccupied scholars from across the humanities and the behavioral and biological sciences, and there are key insights to be found in their work. We can apply these insights about the broader problem of generating cooperative behavior in a mixed population of good- and bad-faith people to the specific problem of how we can build trust and track reality.

Considerable progress on this question of cooperation in a mixed population occurred in the late twentieth century. One spur to progress was the application of a branch of mathematics called game theory, an outgrowth of a field called decision theory. Decision theory addresses the problem of making optimal choices under conditions of uncertainty, whereas game theory addresses the problem of making optimal choices under conditions of conflict, where our choices are interdependent with those of others who have different preferences and incentives. The "games" in game theory are distinct possible combinations of payoffs to each player, contingent on a discrete set of choices each player makes. Game theory examines the performance of different strategies, defined as choices to be made under different contingencies. Though the games take the form of matrices of numbers, they are often given colorful names like "the prisoner's dilemma," "chicken," and "stag hunt."

One of these game structures, "the prisoner's dilemma," is of particular interest because each player is tempted to act selfishly, but if both do so they are worse off than if they had both cooperated. ("The prisoner's dilemma" is named after the scenario in which a prisoner can get a lighter sentence if he squeals on his coconspirator, or they both go free if neither squeals, or they both get a long sentence if they both squeal.) A monetary version of the game, with two players and many rounds of play, has each player privately decide for a given round of the game whether they will cooperate or defect (that is, refuse to cooperate). If both players chose to cooperate, they each get $100, but if one of them has chosen to defect while the other chose to cooperate, the one that defected gets more, $150, and the cooperator gets nothing. The catch is that if neither chose to cooperate, neither gets any money.

Political scientist Robert Axelrod has studied behavior in this sort of game by inviting others to submit strategies for getting the most points in a repeated prisoner's dilemma game in which players had some fixed probability of reencountering each other before the end of the tournament. In the first tournament, the winning strategy was a simple one called Tit for Tat (TFT).[133]

What is TFT? Very simply, the player (1) always cooperates on first encounter with another player, and then (2) simply reciprocates whatever move the other player made last time. Why does TFT thrive? Axelrod characterizes TFT as "nice" (always cooperates on first encounter), "provocable" (it plays less nice against a player that exploited it last time), and "forgivable" (it will cooperate again once the other player starts cooperating). Like a saintly "always cooperate" player, TFT will fare poorly in its first encounter with a selfish player, but after that, it is no longer exploitable. But if it plays another "nice" player, they immediately lock into a profitable pattern of cooperation.

Let's bring this back to the age-old challenge of getting people to work cooperatively on a search for the best possible solution to some real-world problem. This challenge may not map neatly onto any simple game structure like the prisoner's dilemma, but we think some of the lessons of that literature are applicable. There are a lot of temptations that can thwart or abbreviate a cooperative process. But we think something like Axelrod's recipe benefits longer-term cooperative problem-solving. It helps when participants start off by cooperating ("nice") and then are willing to resume cooperating ("forgive") once a recalcitrant partner does so as well. This initial willingness to cooperate seems like a not-so-distant cousin of another trait we've discussed, "scientific optimism"—the ability to hold the belief that a problem can be solved long enough to solve it. Perhaps we need to add a companion concept to our list, "social optimism"—the ability to hold the belief that most people want to cooperate long enough to find cooperative partners and thus solve problems.

But importantly, Axelrod's recipe includes some tough realism: It doesn't pay to continue to cooperate with another player who simply exploits you. For that reason, some cooperative enterprises break down. The key question is whether there are enough cooperative people who can find other cooperative people that these breakdowns don't dominate the story.[134]

In weighing our odds for success, it may be instructive to consider one strategy that surpassed Tit for Tat in later computer competitions. It was the even-more-forgiving "Tit for Two Tats" (TF2T), which wouldn't punish a noncooperating move until it had happened twice. We previously discussed the "fundamental attribution error," the unfortunate tendency in Western, individualistic cultures to assume that others intended a transgression — they are bad actors — whereas when we transgress, it's just an error. Tit for Two Tats acts like a person who is striving to overcome their own tendency to commit the fundamental attribution error. It assumes that the first noncooperating move is just a mistake, not an indicator of a bad-faith actor.

We take hope from TF2T, and we take hope from the real-world accomplishments of those involved in the immense collective efforts that are feeding and educating the world. Our cooperation amid defectors is the source of our best societal, intellectual, and scientific advances. In fact, although the news generally focuses on human conflict, that's a small fraction of our lives. We spend the foundation of our lives in collaborative and cooperative structures. We go to classes where we learn to work with one another and with teachers. We work in businesses where different people have different roles and have to work together for any of them to succeed. We spend our free time with groups of friends.

The psychologist and evolutionary anthropologist Michael Tomasello has for many years argued that humans are unique in the extent of their capacities for collaboration and cooperation.[135] Other animals rarely work together in anything like the scale, range, and variety of ways in which humans cooperate. Cooperation happens very early in our lives: children at nineteen months old will give food they value to a stranger who seems hungry, and will routinely pick up or point to an object that an adult has dropped or mislaid. And collaborative activities, where humans give themselves different roles to play in a social structure with a shared goal, pervade all adult human life. So although we can't depend on each person we encounter being cooperative, at least the odds are better than one might think from reading the news!

THE TELL: A WILLINGNESS TO LEARN

Apparently, then, the first step toward a 3MT-literate, trust-building, reality-tracking culture is social optimism, not because such optimism will be always rewarded, but because it only needs to be rewarded sufficiently while we remain on guard for the bad-faith defectors. In the real world, we do want to know when we have found a good-faith partner, so we can exercise Tit rather than Tat. A good-faith partner would, for example, be committed to finding the truth about factual questions rather than just winning the debate. Thus, the next, decisive step might be our recognition of when we have found other good-faith partners—and perhaps the 3MT willingness to be proven wrong is as good a test as any other, what a poker player might call the tell for a good-faith partner.

We look for this openness to learning in all of our interactions: when we work with other individuals or groups; when we vet possible experts to listen to, and even when we judge institutions—the universities, the newspapers, the professional societies—that provide these experts. So, for example, we listen closely to plausible experts to see how they handle new information that should change their mind, and we look for institutions that support and reward their members when they change their mind. And it is on this basis that we build our next generation of trust networks, both each of us individually and as a society.

As we enter the Third Millennium, this building and rebuilding of trust networks is likely to be front and center in our minds. We have already seen the damages to our society, including extreme political polarization, created by widespread dissemination of false information. And between "clickbait factories" and artificial-intelligence-tuned "sticky" material on the internet and in social media, we can expect to be even more challenged to track reality in the near future. Some existing approaches—person-to-person networks, conferences, universities—can address this challenge, but we have the opportunity (and need) to do more. So we will need to think more practically about a "trust economy" (coined in parallel to the "attention economy") that will find new ways to reward open-minded collective thinking, based on new experiments.

Such experiments could include new mechanisms to support high-quality investigative journalism: For example, there has long been discussion of developing technologies to make it possible for each reader on the internet to pay a very small sum when they read a journalist's work, altogether adding up to substantial pay for good work. Or we may start looking for new practices that encourage demonstrated willingness to be proven wrong: Journalists typically present their stories and analyses as objective and true. Imagine if every news analysis (picture *Economist* articles) ended with a small box in which the writer offered a set of negative "indicators" that readers could watch for in future news; indicators that, if present, would prove the analysis wrong. (For example, "If the unemployment rate rises another point next month, my argument that the new interest rate policy will reverse the current jobs decline will be incorrect.") Not only would this guide the reader in "considering the alternative," but the journalists would themselves be forced to consider the possibility that their analysis will be proven wrong, opening up a more reality-tracking goal rather than just appearing to be smart pundits.

Taking this trust economy experimentation a step further—after all, we have suggested that a 3MT-based society is an "experimenting society"—we will want to invent incentive systems to encourage all the megamedia of our time to optimize positive, mutual learning among the population they are selling to. If the profits of a media company's shareholders depend on nothing but the public's attention and consequent advertising revenues, then there is a natural tendency for a media industry to at best put noise into our mutual public learning and conversation, and at worst drive us into information silos and extremism. This is an information version of a "tragedy of the commons," where polluting the public space is in a company's individual interest but nobody's collective interest. And just as we eventually learned to provide incentives (and penalties) to reform environmental pollution, we will need to experiment with incentives and penalties to reform the pollution of our cognitive commons.

For example, one could imagine developing incentives for social and news media that depended on the degree to which users of a given news medium were led into echo chambers of political thought or, alternatively, were presented with a broad understanding of the political debates. A running survey of every news medium's most stalwart viewers could track whether they were, over time, more or less able to describe the arguments that they disagree with for a randomly selected topic of the day. Businesses would lose the incentives if their users were less able to articulate alternative views, and would benefit from the incentives if their users were better able to. This would provide a brake on polarizing and siloing algorithms of engagement.[136]

But while we are waiting for these larger societal efforts on the cognitive commons to help us, we will also have to experiment individually to build our own trust networks. How many of us have a contrarian friend to talk to, who can give plausible arguments against the positions that most of our friends take for granted? Trust networks are not the same as echo chambers full of like-minded friends; you need to be connected to people with whom you differ, but with whom you can have a real conversation about those differences. If we want to track reality in a world of plentiful but undifferentiated good and bad information, we may have to actively find such friends and contacts. (Perhaps we should set up a match service for readers of this book who might like to find others who will disagree with them, but who share a 3MT passion for tracking reality.)

At the beginning of this chapter we restated the challenge we face as we enter the Third Millennium, the challenge that necessitates Third Millennium Thinking: *With all of our direct access to vast universes of data, we are now forced to figure out what facts to base our decisions on, when to do research for ourselves and when to look for experts, what experts are trustworthy (and on which particular topics), and when we might need wise guidance on integrating values.* But to refine the issue one step further, the challenge isn't just that there's a vast mass of information to sift through; it's that so many of the sources of information boldly claim to be complete and correct. So we need 3MT tools to help us establish the effective trust relationships that can build up a network of credible sources—individuals, experts, institutions, websites—and in the end have the capability of assessing the credibility of conflicting claims.

This is more of a process of construction than an exercise in sifting. When we identify the pieces that are trustworthy, we don't believe in them because our preferred political or cultural group believes in them and the other side doesn't; rather, we believe in them because self-questioning people we disagree with believe in them, too. That's where we build our understanding from.

If we can build on the best of these many developments, what we see emerging—and what we want to highlight and to nurture with Third Millennium Thinking—isn't just a rejuvenation of the Enlightenment; it may turn out to be a second and genuinely new kind of enlightenment for a new millennium.

We hope to leave the reader with a realistic, but exhilarating, sense that with a collective push, we just might be able to turn this corner into the Third Millennium with a new collaborative approach to our problems and opportunities, from the small to the global. We don't need to believe in the vision of an all-flourishing, utopian-sounding world to make this project interesting—there is so much more that is reachable and immediate for us to gain—but maybe some of us will be motivated and excited by that grander goal, too. We have the 3MT ingredients, we have the motivation, and we have reasonable scientific optimism that could make this the millennium in which we advance as a global human family.

We face enormous challenges in the coming decades. But let's remember that the last two decades constitute just 2 percent of our new millennium. The other 98 percent is for us to create.

Acknowledgments

We have many contributions to acknowledge, in the making of this book. First, there are the students who met (over nine months!) during the gestation period for the related university course, and then the dedicated, creative sets of teaching assistants, both graduate and undergraduate, and postdoctoral researchers and consultants, all of whom played active roles in identifying the ideas and topics that we were assembling and how to teach them and assess the results. Over the nine times (so far) that we have taught the course it has been a labor of love for so many of our teaching staff and the course would not be the same without their creative contributions. We wish we could call out important individual efforts, but at least we can warmly thank Adhiraj Ahuja, Ingrid Altunin, Shrihan Argawal, Sophia Baginski, Kasia Baranek, Kristin Barker, Jennifer Barnes, Grant Belsterling, Kelly Billings, Colette Brown, Micah Brush, Jasmine Casey, Paul Christiano, Ethan Chiang, Giana Cirolia, Andrew Critch, Matthew Davis, Brian Delahunty, Ada Do, Moulay Draidia, Katherine Eddinger, Amy Fingerle, Drey Gerger, Tom Gilbert, Leah Gulyas, Nora Harhen, Chad Harper, Quian He, Jacob Heisler, Andrea Hengaertner, Rachel Hood, Rebecca Hu, Christina Ismailos, Kristen Isom, Colin Jacobs, Amisha Jain, Rachel Jansen, Darren Kahan, Louis Kang, Dan Keys, Namrata Khantamneini, Tarah Kirnan, Hannah Laqueur, Alyssa Li, Guang-Chen Li, Emily Liquin, Hui-Chen (Betty) Liu, Ana Lyons, Nina Maryn, Smriti Mehta, Dylan Moore, Nikolai Oh, Gufran Pathan, Antonia Peacocke, Jonathan Pober, Keven Quach, Radhika Rawat, Erin Redwing, Jem Ruf, Trevor Schnack, Vincent Sheu, Riordon Smith, Sophia Steffens, Bethany Suter, Aaron Szasz, Kaitlan

Tseng, Bridget Vaughan, Dax Vivid, Sophie Wiener, Liz Wildenhain, Daniel Wilkenfeld, Alice Zhang, Ted Zhang, Rebecca Zhu, and Zachary Zimmerman. We also thank the nine cohorts of undergraduate students who have taken the course; we have learned a lot from them.

Aditya Ranganathan, Winston Yin, and Gabriel Perko-Engel deserve a very special mention here; over different years, they led our student teaching staff, coordinated our development of teaching materials, and contributed thoughtful content—and enthusiasm—to the course, and this book. Dr. S. Emlen Metz has been a leading force in this work, both in the development of learning goals and assessment materials, and then as a leader and thought partner in developing articulations of the course material as it morphs into different media and targets different audiences. Prof. Alicia Alonzo, as a science educator par excellence, was our earliest guide to such rigorous and testable educational approaches, and her collaboration with us was a large part of the early articulation of the course and its assessment techniques. And after Rob MacCoun moved institutions, we have had several years of wonderful shared teaching with Tania Lambrozo, professor of cognitive science, Alison Gopnik, professor of psychology, and Amy Lerman, professor of public policy and political science, who over different years took over Rob's social science role and brought new ideas and excitement to the course. Johann Frick made fascinating contributions the year he taught in John Campbell's philosophy role. This book captures the story that we see today, but we can expect future sequels!

Will Lippincott provided strong guidance and encouragement as an agent and a thinker. This book would probably remain unwritten if not for Nicole Pagano, who not only helped us organize our collaboration but also was a constant source of wisdom, diplomacy, and enthusiasm as she reacted to each new idea for the book. Lisa Kaufman read early drafts and provided invaluable feedback on how to make the book more accessible and understandable. Eric Engles helped take lecture transcripts and turn them into a more readable form. And we thank Jeffrey Ptak (at Morningstar), Rob Vishny, and Steve Kaplan for their help developing the graph and citations concerning managed mutual funds. We appreciated the excellent team at Little, Brown Spark, led by publisher and editor-in-chief Tracy Behar, who played the key role of the person who grokked our vision for the book and made it happen. A particularly important background contribution was that of Alix Schwartz, who developed and ran the "Big Ideas" course program at UC Berkeley, encouraging this unusual course concept to flourish, and starting us on the path to this book with conceptual and practical support. The Gordon and Betty Moore Foundation supported much work for this book, with Janet Coffey providing expertise and guidance on the book's content and educational approaches that was just as valuable as the funding she oversaw. Karen and Frank Dabby have been longtime thoughtful benefactors. In recent years, the course has also benefited dramatically from the philanthropic support of Mark Rosenthal.

Finally, we thank our families and friends and mentors who participated in this work in multiple roles. Saul learned and discussed many of the concepts in this book with Rich Muller, Bob Cahn, and his father, Daniel Perlmutter. His mother, Felice (Faigie) Perlmutter would have recognized her warm collaborative approach in the book, too. Saul typically runs most of his ideas—and his sentences—by his wife, Laura Nelson; his daughter, Noa Perlmutter, helped him with editing, and designed all the chapter-heading icons; both make a warm, full life. John would like to thank Cassandra Chen, Antonia Peacocke, Niko Kolodny, Tim Crockett, and his son Rory, who all helped in different ways. Rob would like to thank his father, Malcolm MacCoun; his late wife, Lori Dair, who was his pillar of support and fount of wisdom for 35 years before succumbing to ALS in 2022; and his daughters Audrey and Maddie for helping to care for Lori during her illness and then care for him during his battle with cancer in 2022–2023.

In our hope for a future where people listen to each other, propose ideas, and enjoy creating something together, we can't not mention the people who have enriched our lives by sharing music with us. Some of us have taken part in chamber music or orchestras; others play in jazz groups. The joy and skill of engaging in collective music-making is a deep source of the spirit of this book.

Notes

Introduction

1 The university course is called "Sense and Sensibility and Science," and the course material is available for faculty and students at sensesensibilityscience.berkeley.edu.

A high school–level version of the course, "Scientific Thinking for All: A Toolkit," is also being developed and distributed in collaboration with the Nobel Prize Foundation's Outreach division. See nobelprize.org/scientific-thinking-for-all/.

2 National Opinion Research Center (2023, June 15). *Major declines in the public's confidence in science in the wake of the pandemic.* We note, however, that science is still ahead of other professions, for which trust is often falling faster still.

3 What do we mean by "science" in this book? Wikipedia's definition, "Science is a rigorous, systematic endeavor that builds and organizes knowledge in the form of testable explanations and predictions about the universe," captures much of what we are describing, perhaps best supplemented by *Merriam-Webster*'s definition, "Knowledge or a system of knowledge covering general truths or the operation of general laws especially as obtained and tested through scientific method."

Chapter 1: Decisions, Decisions, Decisions

4 Epistocracy is discussed in depth in Estlund, David M. (2009), *Democratic authority,* Princeton University Press. See also Brennan, Jason (2016), *Against democracy,* Princeton University Press.

Chapter 2: Instruments and Reality

5 It's worth noting that these studies of the cognitive effects of indoor air quality are difficult to do well, and there are many variants of measurement techniques both for the properties of the air that the subjects are actually exposed to and the particular cognitive capacities that are potentially affected. There is also the possibility that the diminished cognitive capabilities may be due to the other pollutants building up in the air in a closed room and that the carbon dioxide fraction is here acting as a proxy measure of these other pollutants. For a discussion of all of these issues, see Du, B., Tandoc, M. C., Mack, M. L., & Siegel, J. A. (2020), Indoor CO_2 concentrations and cognitive function: A critical review, *Indoor Air, 30:*1067–1082; and the recent review of Fana, Y., Caoa, X., Zhang, J., Laid, D., & Panga, L. (2023), Short-term exposure to indoor carbon dioxide and cognitive task performance: A systematic review and meta-analysis, *Building and Environment, 237,* 110331.

6 Ronchi, V. (1967.) The influence of the early development of optics on science and philosophy. In E. McMullin (Ed.), *Galileo: Man of science.* Basic Books, 195–206.

7 Heyerdahl, T. (2013). *Kon Tiki*. Simon & Schuster. For the application of the raft and the pyramid metaphors to our knowledge, see Sosa, E. (1980), The raft and the pyramid, *Midwest Studies in Philosophy, 5*, 3–26.

Chapter 3: Making Things Happen

8 The actual story of this correlation is a little hard to figure out (and there are some debated indications that smaller amounts of alcohol may have the opposite correlation for some people). See Godos, J., Giampieri, F., Chisari, E., Micek, A., Paladino, N., Forbes-Hernández, T. Y., Quiles, J. L., Battino, M., La Vignera, S., Musumeci, G., & Grosso, G. (2022). Alcohol consumption, bone mineral density, and risk of osteoporotic fractures: A dose-response meta-analysis. *International Journal of Environmental Research and Public Health, 19*(3), 1515.

9 Pouresmaeili, F., Kamalidehghan, B., Kamarehei, M., & Goh, Y. M. (2018). A comprehensive overview on osteoporosis and its risk factors. *Therapeutics and Clinical Risk Management, 14*, 2029–2049.

10 Sober, E. (2001). Venetian sea levels, British bread prices and the principle of the common cause. *British Journal for the Philosophy of Science, 52*, 331–346.

11 The Spurious Correlations website is using the word "spurious" for cases in which there's simply *no* causal network underpinning the association between two variables, that is, when the association is just a chance coincidence. However, you will often find the term "spurious correlation" used to describe the case where there's a common cause of the two variables, which we previously called Model D.

12 In the discussion that follows, we draw heavily on Judea Pearl's important theory of causality, as well as the work of statistician Donald Rubin and philosopher James Woodward. An accessible introduction to Pearl's account is his 2020 book, *The book of why: The new science of cause and effect*, Basic Books. A more complete and rigorous treatment is his 2009 book, *Causality* (2nd ed.), Cambridge University Press. Rubin's framework appears in Imbens, G. W., & Rubin, D. B. (2015), *Causal inference for statistics, social, & biomedical sciences: An introduction*, Cambridge University Press. Woodward's treatment of the philosophy of intervention can be found in "Causation and manipulability," his 2016 entry in the *Stanford encyclopedia of philosophy*. https://plato.stanford.edu/ENTRIES/causation-mani/

13 It is worth noting two features of "random assignment." First, it is striking that something that seems so uninformative—adding randomness to our study—in fact makes it *more* informative. Second, random assignment is a different procedure than the similarly named "random selection." Random selection is an important way to make our sample results generalizable to a larger population, but it isn't a method for establishing causation, which is a different goal.

14 For example, in the 1960s Dr. Chester M. Southam ran several experiments that involved injecting viruses or cancer cells into human subjects with limited ability to consent to participate in the research. Plumb, R. K. (1964, March 22). Scientists split on cancer tests; some back use of humans—more humility urged. *The New York Times*, 53.

15 Hill, A. B. (1965). The environment and disease: association or causation? *Proceedings of the Royal Society of Medicine, 58*(5): 295–300.

16 In any given application, this kind of nonexperimental evidence can be assessed using the previously cited causal frameworks of Judea Pearl or Daniel Rubin, which enables researchers to clarify which potential causes can be ruled out by the data and which causal explanations remain viable.

Chapter 4: A Radical Shift to Probabilistic Thinking

17 https://www.usgs.gov/faqs/what-probability-earthquake-will-occur-los-angeles-area-san-francisco-bay-area?qt-news_science_products=0#qt-news_science_products, accessed on 2021-11-21.

18 This quote appears in Nicholas Taleb's *Fooled by randomness*, introduced by the words "The Scots philosopher David Hume posed the issue in the following way (as rephrased in the now famous black swan problem by John Stuart Mill)."

19 Reddy, V. (2007). Getting back to the rough ground: deception and "social living." *Philosophical Transactions of the Royal Society, London, B: Biological Science, 362*(1480): 621–637.

20 The journal article that showed evidence for a magnetic monopole discovery was: Price, P. B., Shirk, E. K., Osborne, W. Z., & Pinsky, L.S. (1975), Evidence for detection of a moving magnetic monopole, *Physical Review Letters, 35,* 487. The journal article that said that the scientists had changed their mind was: Price, P. B., Shirk, E. K., Osborne, W. Z., & Pinsky, L. S. (1978), Further measurements and reassessment of the magnetic-monopole candidate, *Physical Review D, 18,* 1382.

Chapter 5: Overconfidence and Humility

21 The Chernobyl and Challenger cases are discussed in Freudenburg, W. R. (1988), Perceived risk, real risk: Social science and the art of probabilistic risk assessment, *Science, 242,* 44–49.

22 A 2012 study of more than two thousand scientific retractions found that less than a quarter were attributed to an error; two-thirds were attributed to misconduct: Fang, F. C., Steen, R. G., & Casadevall, A., (2012), Misconduct accounts for the majority of retracted scientific publications, Proceedings of the National Academy of Sciences, USA, 109, 17028–17033. (In an interesting twist, the authors later published a correction based on errors in a table in their paper.) This study might give the impression that scientific misconduct is far more common than scientific error—but surely the opposite is true. In fact, other studies show that experts are sometimes reluctant to acknowledge errors, and when they do, they often do so in a way that minimizes the errors and the threat they pose. (See Tetlock, P. E. [2006]. *Expert political judgment: How good is it? How can we know?* Princeton University Press.)

23 Asness, C., et al. [23 authors] (2010, Nov. 15). Open letter to Ben Bernanke. *The Wall Street Journal.* Carey, D., & Willmer, S. (2014, Oct. 10). Fed naysayers warning of inflation say they're still right. *Bloomberg.*

24 Krugman, P. (2022, Jan. 21). Honey, I shrank the economy's capacity. *The New York Times.*

25 Leary, M. R., (2018), *The psychology of intellectual humility,* John Templeton Foundation. https://www.templeton.org/wp-content/uploads/2020/08/JTF_Intellectual_Humility_final.pdf

26 Rohrer, J. M., Tierney, W., Uhlmann, E. L., DeBruine, L. M., Heyman, T., et al. Putting the self in self-correction: Findings from the Loss-of-Confidence Project. *Perspectives on Psychological Science, 16:* 1255–1269.

27 This task was developed by Koriat, A., Lichtenstein, S., & Fischhoff, B. (1980), Reasons for confidence, *Journal of Experimental Psychology: Human Learning and Memory, 6,* 107–118. We caution that for technical reasons this particular task can overstate the magnitude of overconfidence, but other (more complex) methods show that the effect is real. See Moore, D. A., & Healy, P. J. (2008). The trouble with overconfidence. *Psychological Review, 115,* 502–517.

28 Researchers have proposed alternative explanations for the typical calibration pattern in this kind of "2-alternative forced choice" task; see Koriat, A. (2012), The self-consistency model of

subjective confidence, *Psychological Review, 119,* 80–113. But the general finding of overconfidence has been widely replicated using many other procedures.

29 A confidence interval is one form of the "error bar" discussed in chapter 4.

30 Deaves, R., Lüders, E., & Schröeder, M. (2010). The dynamics of overconfidence: Evidence from stock market forecasters. *Journal of Economic Behavior & Organization, 75,* 402–412.

31 Tetlock, P. E. (2006). *Expert political judgment: How good is it? How can we know?* Princeton University Press.

32 Birge, R. T. (1941). The general physical constants: As of August 1941 with details on the velocity of light only. *Reports on Progress in Physics, 8,* 90–135.

33 Henrion, M., & Fischhoff, B. (1986). Assessing uncertainty in physical constants. *American Journal of Physics, 54,* 791–798.

34 Murphy, A. H., & Winkler, R. L. (1977). Reliability of subjective probability forecasts of precipitation and temperature. *Journal of the Royal Statistical Society,* Series C (Applied Statistics), *26,* 41–47. We should note that one study found blackjack dealers to be no better calibrated than lay people when judging blackjack choices, even though the dealers presumably have some of the same advantages as meteorologists. See Wagenaar, W., & Keren, G. B. (1985). Calibration of probability assessments by professional blackjack dealers, statistical experts, and lay people. *Organizational Behavior and Human Decision Processes, 36,* 406–416.

35 Wakeman, N. (2011, Feb. 11). IBM's 'Jeopardy!' match more than game playing. *Washington Technology.* https://washingtontechnology.com/articles/2011/02/10/ibm-watson-data-uses.aspx. As it happens, new analyses show that Watson was also slightly overconfident rather than perfectly calibrated. In any case, Watson was still more careful about offering opinions than human players. See Moore, D. (2023), Overprecision is a property of thinking systems, *Psychological Review, 130,* 1339–1350.

36 Wells, G. L., Lindsay, R. C. L., & Ferguson, T. J. (1979). Accuracy, confidence, and juror perceptions in eyewitness identification. *Journal of Applied Psychology, 64,* 440–448.

37 Tenney, E. R., MacCoun, R. J., Spellman, B. A., & Hastie, R. (2007). Calibration trumps confidence as a basis for witness credibility. *Psychological Science, 18,* 46–50. Tenney, E. R., Spellman, B. A., & MacCoun, R. J. (2008). The benefits of knowing what you know (and what you don't): Fact-finders rely on others who are well calibrated. *Journal of Experimental Social Psychology, 44,* 1368–1375.

38 Sah, S., Moore, D., & MacCoun, R. (2013). Cheap talk and credibility: The consequences of confidence and accuracy on advisor credibility and persuasiveness. *Organizational Behavior and Human Decision Processes, 121,* 246–255.

39 www.theguardian.com/books/2015/jul/18/daniel-kahneman-books-interview

Chapter 6: Finding Signal in Noise

40 The data and plots for the global average surface temperature measurements are from Rohde, R. A., & Hausfather, Z., (2020). The Berkeley Earth Land/Ocean Temperature Record, *Earth System Science Data, 12,* 3469–3479.

41 The Greenland ice-core data is from Vinther, B.M., Buchardt, S. L., Clausen, H. B., Dahl-Jensen, D., Johnsen, S. J., et al. (2009), Holocene thinning of the Greenland ice sheet, *Nature, 461,* 385–388.

42 See: Rohde R., Muller R. A., Jacobsen, R., Muller, E., Perlmutter S., et al. (2013). A new estimate of the average earth surface land temperature spanning 1753 to 2011, *Geoinformatics & Geostatistics: An Overview, 1.*

43 It's worthwhile considering the relationship between the concept of signal versus noise and the concept of causality we discussed earlier. When you are trying to determine what causes what, the causal variable is the one you're trying to isolate so you can see its signal clearly in causing the effect you're studying. The other variables are producing phenomena that act as noise because they are irrelevant to the relationship you are looking at but are present in the system (and often make it hard to tell the effect of the causal variable, that is, the signal).

Chapter 7: Seeing Patterns in Random Noise

44 This plot is based on the discovery graph presented in "Latest Results from ATLAS Higgs Search," 4 July 2012, by the ATLAS Collaboration, atlas.cern/updates/press-statement/latest-results-atlas-higgs-search.

45 This plot is based on the discovery graph presented in "CMS Higgs Seminar: Images and plots from the CMS Statement," 4 July 2012, by the CMS Collaboration, cds.cern.ch/record/1459463.

46 For perhaps obvious reasons, this question of whether there is any persistence in the relative success of fund managers gets revisited over the years. A recent analysis again shows that you should not bet money on it: Choi, J. J., & Zhao, K. (2021), Carhart (1997) mutual fund performance persistence disappears out of sample, *Critical Finance Review, 10,* 263–270. cfr.pub

47 Hodis, H. N., & Mack, W. J., (2013), The timing hypothesis and hormone replacement therapy: A paradigm shift in the primary prevention of coronary heart disease in women. Part 1: comparison of therapeutic efficacy, *Journal of the American Geriatrics Society, 61,* 1005–1010; Hochberg, Y., & Westfall, P. H. (2000), On some multiplicity problems and multiple comparison procedures in biostatistics, in P. K. Sen & C. R. Rao (Eds.), *Handbook of statistics, vol. 18,* Elsevier Science, pp. 81–82.

48 At this point, you might start wondering whether it's better to get more data or not. If you are using a data set to hunt for a number of different possible causal factors (the variables of your study), we just said that you'd need to pre-commit to your variables ahead of time (not make them up on the fly) and that more variables will require more data to counterbalance the "look elsewhere" effect. But before that we also said that with more data it's likely you'd see more patterns that look like "signal" amid the noise. So is more data better or worse? And what if you do see a pattern that looks like evidence for something you hadn't anticipated: are you supposed to just ignore it? (Think, for example, of a medical trial where the researchers see that the drug being tested has a curative effect on a disease that was not the target of the drug. What then?)

It is precisely these sorts of questions that get us to think more deeply—and quantitatively—about our probabilistic understanding of signal and noise. First, when we hunt for signal in noise, what we are really doing is comparing how often we see what looks like a signal in the data to how often we would expect to see what looks like a signal if there is no real signal, just spurious patterns occasionally appearing in the noise. Then the advantage of collecting more data is that you can make a better and better prediction of how often you will see spurious noise patterns, and then compare that number to the number of real-plus-spurious patterns that you see in the data. That comparison tells us the probability that what we are seeing is just noise—or, conversely, a real signal amid the noise. This is essentially what most statistics techniques help us to do.

Next, if you do see some evidence for a signal, say, some causal relationship, that you didn't set out to test, this just means that you have set your filter for what counts as interesting to a much more inclusive setting; it's not filtering out as much noise. This leads to a much higher rate of false alarms compared to real signal, so your odds of just seeing spurious noise patterns is higher relative to the real signal. If you want to return to the better odds, you now need to collect still more data. Again, this is what statistics helps us figure out.

49 Instrumentation has a very direct relationship to the issue of finding signal in noise. Instruments are often designed to amplify or magnify things in nature so that we can observe them with our limited sensory abilities. But in the absence of any filtering device, they are indiscriminate about signal and noise. They will amplify both. So the instrumentation may not leave you any better off unless you have an idea of what the signal may be and how some of the noise can be filtered out. Some instruments are designed specifically to offer you options that enable filtering.

Chapter 8: Pick Your Poison": Two Kinds of Error

50 See MacCoun, R. J. (2024.) Standards of proof: Theory and evidence. In R. Hollander-Blumhoff (Ed.), *Research handbook in law and psychology*. Elgar.

51 Arguably, the optimal threshold (call it "p*") should be based on the following formula: p* = Aversion to a False Positive/(Aversion to a False Positive + Aversion to a False Negative) where aversion might range from 0 (indifferent to this error) to 100 (maximum aversion).

52 Everything else being equal, this implies a p* of 10/(10+1) = .91.

53 Even when you find evidence of, say, a new elementary particle, there is a value judgment regarding the announcement in the context of uncertainty, given the number of theorists who may go off on a wild goose chase if the finding proves incorrect.

54 In real life, of course, colleges do not admit everyone who applies one year just to learn what would happen. That would be impractical (most schools have limits on how many students they can accommodate), and it would be cruel (since many admitted students would fail to pass their courses). As a result, real-world institutions rarely get to see the full picture portrayed in these figures, and they are forced to use other information (e.g., performance at whatever college a rejected student ended up attending) as a proxy.

55 Kliff, S., & Bhatia, A. (2022, Jan. 1). When they warn of rare disorders, these prenatal tests are usually wrong. *The New York Times.*

56 Bayes Rule (sometimes described as Bayes Theorem) is relevant to many topics in this book. It tells us that if you want to know what the probability is that some statement is correct, given some new relevant data, you also have to know what we previously knew about the probability that the statement is correct—since usually, you already have some prior indications of a statement's probability! Bayes Rule (really just a formula for our purposes) tells us exactly what the correct updated probability is after you get new data. You can find several different (equivalent) versions of the formula online, but roughly, it says to take your previous best estimate of the probability and revise it by multiplying by the likelihood that you would see the new data if your statement is true divided by the likelihood that you would see the new data in any case whether the statement was true or not.

It is interesting to note that there are lines of research suggesting that people don't always update their probabilities in accord with Bayes theorem. You might think "Well, I know I don't—I've never even seen the formula before." But our brains are surely capable of Bayesian updating. For example, there is evidence that the tiny little brains of bees enable them to engage in "optimal Bayesian foraging," and human inferences do approximate Bayes theorem fairly well on some tasks. But we also sometimes rely on non-Bayesian heuristics or reasoning strategies, presumably because they offer benefits (e.g., speed, ease of use, or ease of communication) that outweigh a reduction in accuracy.

There are plenty of visual introductions to Bayes Theorem on YouTube. For a broad and accessible introduction to Bayesian thinking, see Sharon Bertsch McGrayne's *The theory that would not die: How Bayes' rule cracked the Enigma code, hunted down Russian submarines, & emerged triumphant from two centuries of controversy* (Yale University Press, 2011).

Chapter 9: Statistical and Systematic Uncertainty

57 A given source of noise can act as either a statistical uncertainty or as a systematic uncertainty depending on how it is entering into the measurement that you are making. For example, the miscalibration of a bathroom scale can result in statistical uncertainty if you're trying out a lot of scales (each of which is randomly miscalibrated) or systematic uncertainty if you only use one over and over again.

58 The arrows in the image are courtesy of vecteezy.com/free-vector/dart. The dartboard has been modified from vecteezy.com/free-vector/dart-board.

59 The statisticians may have come up with the terminology for these two types of uncertainty that seems most arbitrary etymologically, since they use "better accuracy" for "better systematic uncertainty" and "better precision" for "better statistical uncertainty."

60 To make matters worse, the random component—the part that *should* be easiest for us to handle—can trick us into seeing patterns that aren't there, as we saw in chapter 7.

61 To address statistical uncertainty, the sample size is determined mathematically by your desired confidence interval, and by how large your acceptable error bars are. For example, in a city of 200,000 people, to learn how many will vote for which mayoral candidate, you would need to poll about 2,400 people to have 95 percent confidence that the actual number is likely to be within 2 percent of the poll result.

Chapter 10: Scientific Optimism

62 Apparently, when experimenters measure the energy used by people thinking hard, it's only somewhat higher than the brain's constant energy use. The brain is, however, a big energy hog, accounting for something like a fifth of our energy use even when we are relaxing, so maybe even a small proportional increase is perceptible. Or perhaps it's that we are often under stressful conditions when we think hard (e.g., for a test at school), and it's the energy we expend on our stress response that we are aware of.

63 For that small fraction of readers who, like Saul, took a physics course in high school and then went on to take a college course, you are likely to recognize this example: After getting used to solving high school physics problems in just a few minutes, it can be a shock to come across college physics problems that can each take several hours of iterative thinking. And if you don't know that you will eventually solve them, there is a tendency to give up too soon, before you discover that you can.

64 These scientific traditions are often handed down from one generation of scientists to the next. In this case, Saul was learning it from his research advisor, Richard Muller, who in turn learned it from his research advisor, the Nobel laureate Luis Alvarez. We should look to see who taught this can-do approach to Luis Alvarez—perhaps it was *his* research advisor, Arthur Compton (also a Nobel laureate). Of course, our hope is to spread this to the much larger audience that may or may not be professional scientists!

65 For this to work, the sources of funding have to understand this point, too. Take note, public science funding agencies!

Chapter 11: Orders of Understanding and Fermi Problems

66 Even the idiom "cooking with gas" suggests the idea of scientific progress. The phrase originated in the late 1930s as a comedy catchphrase at a time when gas stoves had been replacing slower wood-burning stoves. Apparently, the gas industry was thrilled by the catchphrase being used extensively in radio comedy shows.

67 As with many terms used in the physical sciences, "first order understanding of a problem" is jargon from the field of mathematics. If you've had a calculus class, you might recognize it as being related to Taylor Series expansions.

The basic idea is that you can approximate any function at some location, *x*, by first looking at the function, *f*(*a*), at a nearby value, *a*, and then adding the first derivative, *f'*(*a*), times the distance from that value, (*x*—*a*), and then a second derivative times that distance squared, and so on until you have made up a series that will approximate the function. The way that we think about it is that as you add more and more terms into the approximation, you get closer and closer to matching what the function might actually look like. When you're making a first order of approximation of something, all you mean is that you've only used the first term with one derivative in your approximation. And for second order you've used one more term, and for third order you've used another term. Little by little it starts to look more and more like what's really going on in the function you're trying to approximate.

Thus, the first-order explanation for something captures the most obvious salient causal relationship—what's really making this thing happen. The second-order explanation is the first one that captures some subtlety of what's going on, the exceptions or small additions to this first approximation. The third order is what expresses a yet smaller modification to that last modification, and so on. (There can be good reasons that humans sometimes ignore first-order explanations; e.g., arson investigators don't report that the primary cause of a house fire was the presence of oxygen. Sometimes when we want to emphasize that something is the glaringly obvious, dominant cause, we call it the zeroth-order explanation.)

68 The insomnia example also raises the point that for any given phenomenon, what constitutes a lower- versus higher-order factor can vary with the context. For example, if a hungry lion started prowling outside your bedroom, that might quickly become the first-order factor keeping you awake.

69 Policy analysts call these rapid estimates "BOTEC" judgments (back-of-the-envelope calculations).

70 If you are looking for examples to play with, and further discussion of estimation techniques, there are books that focus on just this. Two you might look at are: *Guesstimation: Solving the world's problems on the back of a cocktail napkin,* by Lawrence Weinstein and John A. Adam (Princeton University Press, 2009); and *Maths on the back of an envelope: Clever ways to (roughly) calculate anything,* by Rob Eastaway (HarperCollins, 2019).

Chapter 12: Why It's Hard to Learn from Experience

71 See McDaniel, M. A., Schmidt, F. L., & Hunter, J. E. (1988). Job experience correlates of job performance. *Journal of Applied Psychology, 73,* 327–330; and Dokko, G., Wilk, S. L, & Rothbard, N. P. (2008). Unpacking prior experience: How career history affects job performance. *Organization Science, 20,* 51–68.

72 This issue is discussed in more detail in MacCoun, R. J., (1998), Biases in the interpretation and use of research evidence, *Annual Review of Psychology, 49,* 259–287.
73 https://en.wikipedia.org/wiki/List_of_cognitive_biases
74 Gigerenzer, G., & Goldstein, D. G. (2011). The recognition heuristic: A decade of research. *Judgment and Decision Making, 6,* 100–121.
75 Although Dr. Seuss did come close when he got to the letter X: "X is very useful if your name is Nixie Knox. It also comes in handy, spelling ax and extra fox."
76 Lichtenstein, S., Slovic, P., Fischhoff, B., Layman, M., & Combs, B. (1978). Judged frequency of lethal events. *Journal of Experimental Psychology: Human Learning and Memory, 4,* 551–578.
77 Bailis, D. S., & MacCoun, R. J. (1996). Estimating liability risks with the media as your guide: A content analysis of media coverage of civil litigation. *Law and Human Behavior, 20,* 419–429.
78 See Ellman, I. M., Braver, S., & MacCoun, R. J. (2009). Intuitive lawmaking: The example of child support. *Journal of Empirical Legal Studies, 6,* 69–109.
79 Fischhoff, B. (1975). Hindsight is not equal to foresight: The effect of outcome knowledge on judgment under uncertainty. *Journal of Experimental Psychology: Human Perception and Performance, 1,* 288–299.
80 Tajfel, H., Flament, C., Billig, M. G., & Bundy, R. P. (1971). Social categorization and intergroup behavior. *European Journal of Social Psychology, 1,* 149–177.
81 "By fusing contending positions on a risk or like facts to opposing group identities, antagonistic memes effectively transform positions on them into badges of membership in, and loyalty to, competing groups." Kahan, D. M., Jamieson, K. H., Landrum, A., & Winneg, K. (2017). Culturally antagonistic memes and the Zika virus: An experimental test. *Journal of Risk Research, 20,* 1–40.
82 MacCoun, R. (1993). Blaming others to a fault? *Chance, 6,* 18, 31–33.
83 See Ross, L. D. (1977). The intuitive psychologist and his shortcomings. In L. Berkowitz (Ed.), *Advances in Experimental Social Psychology* (vol. 10, pp. 174–220). Academic Press.
84 Menon, T., Morris, M. W., & Chiu, C. (1999). Culture and the construal of agency: Attribution to individual versus group dispositions. *Journal of Personality and Social Psychology, 76,* 701–717.
85 Merton, T. (1965). *The way of Chuang Tzu,* chapter 20, published by New Directions.
86 Lord, C. G., Lepper, M. R., & Preston, E. (1984). Considering the opposite: A corrective strategy for social judgment. *Journal of Personality and Social Psychology, 47,* 1231–1243.

Chapter 13: Science Gone Wrong

87 The recorded lecture by Irving Langmuir from 1953 was subsequently transcribed and edited by Robert N. Hall as a 1966 General Electric Laboratory report. This report was often copied and passed among scientists, until much later when it was formally published: Langmuir, I., & Hall, R. N. (1989), Pathological science, *Physics Today, 42,* 36-48.
88 Langmuir's definition is clearly related to the "disconfirmation bias" we note in chapter 14—a corollary of confirmation bias in which people only search for flaws in evidence when it seems to disconfirm their hypothesis.
89 Fanelli, D. (2009). How many scientists fabricate and falsify research? A systematic review and meta-analysis of survey data. *PLoS ONE, 4:* e5738.
90 Oddly enough, many papers are caught faking their results by using a photograph that shows results from another, different, already-published scientific study by the same author and

claiming it's from the current study. Of course, this is classic arms race: As scientists learn how to detect fraud, other scientists will develop more subtle ways of committing fraud, which will work until those frauds are detectable... and so on.

91 There is apparently a whole category of words that well-educated people will have learned from reading but don't recognize as a word they have heard pronounced, so they (mis)pronounce them as they appear to them; "epitome" is a classic example. In a book focusing on the many ways that we tend to fool ourselves, we can't use the word "misled" without a comment about its apparent propensity to be mispronounced as the past tense of the nonexistent verb, "to misle." (Whether this verb itself is to be pronounced "mizz-le" or "my-zle" is unclear.) Some writers discussing this phenomenon have even used "misle" as a term to describe such mispronounced words.

92 Langone, J. (1988, Aug. 8). Science: The water that lost its memory. *Time.*

93 The Editors (1988). When to believe the unbelievable. *Nature, 333,* 787.

94 Maddox, J., Randi, J., & Stewart, W., (1988). "High-dilution" experiments a delusion. *Nature, 334,* 287–290.

95 Goldacre, B. (2007, Nov. 17). Benefits and risks of homoeopathy. *Lancet, 370,* 9600, pp. 1672–1673. Note that the "benefits" mentioned were only those obtained with any placebo (dummy pill).

96 Stolberg, M. (2006 Dec.). Inventing the randomized double-blind trial: The Nuremberg salt test of 1835. *Journal of the Royal Society of Medicine, 99*(12): 642–643.

97 It is, however, sobering to note that, before the Nazi period, Germany had developed some of the most advanced policies to protect individuals from medical experimental exploitation. This suggests that, in addition to protocols, structures of oversight are essential to prevent abuse.

Chapter 14: Confirmation Bias and Blind Analysis

98 Wason, P. C., & Johnson-Laird, P. N. (1972). *The psychology of reasoning: Structure and content.* Harvard University Press.

99 Edwards, K., & Smith, E. E. (1996). A disconfirmation bias in the evaluation of arguments. *Journal of Personality and Social Psychology, 7,* 5–241.

100 Okay, if you really want to know, the universe's expansion rate measurement—that is, the measurement of what is called the Hubble Constant—depends on how far apart two points are in the universe that you are watching separate from each other. If they are currently 2 billion miles apart, they move farther apart from each other at about twice the speed that they would if they were currently just 1 billion miles apart. So we measure the expansion rate of the universe in units of speed (km/sec) per distance ("megaparsec"—which is about 2×10^{19} miles).

101 See Klein, J. R., & Roodman, A. (2005). Blind analysis in nuclear and particle physics. *Annual Review of Nuclear and Particle Physics, 55,* 141–163. Also see MacCoun, R., & Perlmutter, S. (2015). Hide results to seek the truth. *Nature, 526,* 187–189.

102 Examples of other uses of blinding include professors grading student work with only the student identification numbers, not students' names; and applications for time on the Hubble Space Telescope are reviewed without the scientists' names.

103 See MacCoun, R. J. (2020). Blinding to remove biases in science and society. In R. Hertwig & C. Engel (Eds.), *Deliberate ignorance: Choosing not to know.* MIT Press.

104 Committee on Identifying the Needs of the Forensic Sciences Community (2009). *Strengthening forensic science in the United States: A path forward.* National Research Council, National Academies Press; President's Council of Advisors on Science & Technology (2016). *Forensic*

science in criminal courts: Ensuring scientific validity of feature-comparison methods. Report to the President, Executive Office of the President.

105 Dror, E., Charlton, D., & Péron, A. E. (2006). Contextual information renders experts vulnerable to making erroneous identifications. *Forensic Science International, 156,* 1, pp. 74–78.

106 For example, Kaiser advises patients to "have your first-opinion records sent ahead to the other doctor." https://healthy.kaiserpermanente.org/health-wellness/health-encyclopedia/he.getting-a-second-opinion.ug5094

107 For overviews of the Open Science Movement, see Munafò, M. R., Nosek, B. A., Bishop, D. V. M., Button, K. S., Chambers, C. D., et al. (2017), A manifesto for reproducible science, *Nature Human Behaviour, 1,* 1–9; and Jussim, L., Stevens, S. T., & Krosnick, J. A. (Eds.) (2022), *Research integrity in the behavioral sciences* (pp. 295–315), Oxford University Press.

108 For an early example, see Latham, G. P., Erez, M., & Locke, E. A. (1988), Resolving scientific disputes by the joint design of crucial experiments by the antagonists: Application to the Erez–Latham dispute regarding participation in goal setting, *Journal of Applied Psychology, 73,* 753–772. A recent example is Melloni, L., et al., (2023), An adversarial collaboration protocol for testing contrasting predictions of global neuronal workspace and integrated information theory, *PLoS ONE, 18:* e0268577.

109 Note that this is a very different use of the term "badges"—a laudatory one—which should be distinguished from the "badging" phenomenon we refer to elsewhere in the book, where people adopt positions on an issue simply to advertise their commitments to their political party or other groups.

110 See the Rohrer et al. (2021) citation in chapter 5's endnotes, describing the Loss-of-Confidence Project.

Chapter 15: The Wisdom and Madness of Crowds

111 On deindividuation, see Postmes, T., & Spears, R. (1998), Deindividuation and antinormative behavior: A meta-analysis, *Psychological Bulletin, 123,* 238–259. On emotional contagion, see Herrando, C., & Constantinides, E. (2021), Emotional contagion: A brief overview and future directions, *Frontiers in Psychology, 12,* Article 712606.

112 Rob had an opportunity to implement some of Janis's ideas for preventing groupthink while working on a RAND project for the Department of Defense in 1993. RAND had been commissioned to examine whether the military could lift the ban on service by openly gay or lesbian personnel. At the time, this was a highly polarizing topic, and the RAND team tried hard to be scrupulous in showing that it was being impartial and unbiased. After several days of internal briefings, in which they reviewed all the evidence they had compiled, they formed a set of subgroups. Each subgroup contained a mix of military and nonmilitary researchers trained in various academic disciplines (including law, medicine, organizational behavior, social psychology, economics, anthropology). Each subgroup then spent a day reviewing all the evidence and reaching a consensus on whether the evidence suggested that lifting the ban would impair military unit performance. Each group independently reached the same answer: No. President Clinton was briefed on this conclusion, but he did not lift the ban, instead adopting the "Don't ask, don't tell" compromise. The RAND team was reassembled some years later to help President Obama revisit the issue, and the ban was ultimately lifted with remarkably little drama and no subsequent evidence of harm to military effectiveness.

113 See Laughlin, P. R. (2011). *Collective induction.* Princeton University Press.

114 This "social decision scheme" approach is outlined by Stasser, G., Kerr, N. L., & Davis, J. H. (1989), Influence processes and consensus models in decision-making groups. In P. B. Paulus (Ed.), *Psychology of group influence* (pp. 279–326), Lawrence Erlbaum Associates. Note that the simple majority and truth-wins processes can be seen as special cases of a more general logistic threshold model of group influence; see MacCoun, R. J. (2012), The burden of social proof: Shared thresholds and social influence, *Psychological Review, 119,* 345–372.

115 It is probably more accurate to put scare quotes around the word "truth" when discussing truth-wins processes, because the conceptual systems shared by group members wouldn't necessarily be accepted by others outside the group.

116 Kerr, N., MacCoun, R. J., & Kramer, G. (1996). Bias in judgment: Comparing individuals and groups. *Psychological Review, 103,* 687–719.

117 Scott Page has published theoretical and empirical analyses that elaborate on this point. He provides an accessible introduction in his 2007 book, *The difference: How the power of diversity creates better groups, firms, schools, and societies,* Princeton University Press.

Chapter 16: Weaving Facts and Values

118 See Hammond, K. R., & Adelman, L. (1976). Science, values, and human judgment. *Science, 194,* 389–396.

119 See MacCoun, R., Reuter, P., & Schelling, T. (1996). Assessing alternative drug control regimes. *Journal of Policy Analysis and Management, 15,* 1–23; and MacCoun, R., & Reuter, P. (2001). *Drug war heresies: Learning from other vices, times, and places.* Cambridge University Press.

120 Schwartz, S. H. (1992). Universals in the content and structure of values: Theoretical advances and empirical tests in 20 countries. *Advances in Experimental Social Psychology, 25,* 1–65. There are many other similar schemes (including a more recent one by Schwartz), but the Schwartz system appears to provide the best statistical description of survey data.

121 Tetlock, P. E., Peterson, R. S., & Lerner, J. S. (1996). Revising the value pluralism model: Incorporating social content and context postulates. In C. Seligman, J. M. Olson, & M. P. Zanna, *The psychology of values: The Ontario symposium* (vol. 8). Lawrence Erlbaum Associates.

122 Steele, C. M. (1988). The psychology of self-affirmation: Sustaining the integrity of the self. *Advances in Experimental Social Psychology, 21,* 261–302.

123 Sherman, D. K. (2013). Self-affirmation: Understanding the effects. *Social Psychology and Personality Compass, 7,* 834–845.

124 The term comes from Rawls, J., (1971), *A theory of justice,* Harvard University Press.

125 Strawson, P. F. (1980). Review of Ryle, G., On thinking. *Mind 30,* 365–367.

Chapter 17: The Deliberation Challenge

126 For a dramatic example of this desire to appear knowledgeable and have an opinion, see: Bishop, G., Oldendick, R., Tuchfarber, A., & Bennett, S. (1980), Pseudo-opinions on public affairs, *Public Opinion Quarterly, 44,* 198-209. Bishop and his colleagues ran a poll in which they asked people their view on "the repeal of the 1975 Public Affairs Act." Even though there is no such "Public Affairs Act," a third of the public surveyed stated an opinion on this topic.

127 The best introduction to scenario planning is probably Schwartz's 1996 book, *The art of the long view* (Currency). Newer developments and variations are often reported in the journal *Futures,*

including comprehensive literature reviews by Varum and Melo in 2009, and Amer and colleagues in 2013.

128 Tetlock, P. E., Mellers, B. A., Rohrbaugh, N., & Chen, E. (2014). Forecasting tournaments: Tools for increasing transparency and improving the quality of debate. *Current Directions in Psychological Science, 23,* 290.

129 The Public Editor project began with conversations between Saul and Nicholas Brigham Adams (then at "BIDS," the Berkeley Institute for Data Science), who had been developing crowdsourcing techniques and software that enabled group annotation of texts. Adams set up the Goodly Lab nonprofit to make projects like Public Editor possible and has been leading this effort, in collaboration with Saul at BIDS. The current website can be found at publiceditor.io.

Chapter 18: Rebooting Trust for a New Millennium

130 In the late twentieth century, an intellectual movement known as post-modernism rightly called attention to the ways in which the authority of science was used as a kind of Trojan horse to sneakily advance the interests of the privileged and powerful. But it painted a portrait of scientists as naive adherents to a dogma in which science is made of uncontestable "facts" linked together by mathematics using the machinery of proof provided by deductive logic. As far as we could tell, that portrait doesn't square with the actual ways in which working scientists talk about their work. Scientists recognize "proof" as the backbone of logic and mathematics but are painfully aware that empirical science is a web of fallible and provisional observations, constantly subject to revision. By the dawn of the twenty-first century, it mostly seemed as if post-modernism had run its course. But to our astonishment, as the academic post-modernism of the left was winding down, a new form of populist post-modernism emerged outside of academia, on the right. Suddenly, new "facts" could be asserted without hard evidence, and any claims to the contrary could be readily dismissed as "fake news." Although the post-modernists would doubtless be dismayed to note the connections, populism seems to exemplify just the kind of approach they described as the only one: the idea of "objective truth" goes out the window, and hypotheses are assessed only in the light of group ideologies.

At times we have despaired of these developments, and the harms that they have caused to public discourse and shared public problem-solving. But in the end, we don't think populist post-modernism is any more sustainable than its academic predecessor. Ultimately, people want to solve real problems in the real world, and they can do so only through the hard work of sifting through fallible evidence and provisional hypotheses in a search for reproducible outcomes that provide tangible benefits and improvements in their lives. The tools and attitudes that are emerging in the Third Millennium of science reject any notion of a scientific priesthood with a monopoly on "Truth"; there is a shift toward decentralized authority, aggressive fact-checking, and citizen engagement. But these tools can't take hold if they are deployed only by academic scientists; they require the engagement of the whole community. And that is why we have written this book.

131 Originally written in 1971, this essay has been updated several times. See Dunn, W. N. (Ed.) (1998). *The experimenting society: essays in honor of Donald T. Campbell,* published by Transaction Publishers.

132 Our notion of habits of community owes a big debt to the seminal sociologist of science, Robert K. Merton, who articulated four aspirational norms of science (forming the acronym "CUDOS"): *Communism* (scientific knowledge should belong to everyone); *Universalism*

(truth must be judged by impersonal criteria); *Disinterestedness* (personal self-interest should play no role in scientific inquiry); and *Organized Skepticism* (the scientific community should rigorously scrutinize all scientific claims). See his 1973 book, *The sociology of science: Theoretical and empirical investigations* (University of Chicago Press). Earlier chapters of our book (and a great deal of social science work on the practice of science) illustrate how scientists fulfill (and often fail to fulfill) these aspirations.

133 See Axelrod, R. (1984). *The evolution of cooperation*. Basic Books. The kind of reciprocity Axelrod examines is just one of many mechanisms that support (or thwart) human cooperation; for a broader review, see Heinrich, J., & Muthukrishna, M. (2021), The origins and psychology of human cooperation, *Annual Review of Psychology, 72,* 207–240.

134 Of course, any attempt at collaborative decision-making depends on the willingness of people to settle matters by discussion. Sometimes, partisan conflict can reach a pitch where that no longer seems possible. If you propose a deliberative poll to your opponents, and they respond with violence, how do you proceed? Our reflections here are all in the service of making truly joint decision-making happen, and staving off that moment of primal conflict.

135 See, e.g., Tomasello, M. (2009.) *Why we cooperate.* MIT Press.

136 The real-world realization of such incentives isn't as unlikely as it might seem. For example, the Digital Services Act of the European Commission has already put in place an auditing regime for major online platforms that could adopt the survey concept we suggest. More generally, behavioral economists and game theorists have developed many insights into how to design auctions and other procedures to promote sincere bidding and honest information exchange. For a review, see Haaland, I., Roth, C., Wohlfart, J. (2023), Designing information provision experiments, *Journal of Economic Literature 61*, 3–40.

Index

T

About the Authors

Saul Perlmutter is a 2011 Nobel laureate, sharing the prize in physics for the discovery of the accelerating expansion of the universe. He is a professor of physics at the University of California, Berkeley, and a senior scientist at Lawrence Berkeley National Laboratory. In addition to other awards and honors, he is a member of the National Academy of Sciences, the American Academy of Arts and Sciences, and the American Philosophical Society. Perlmutter has written popular articles and appeared in documentaries for PBS, Discovery Channel, and the BBC.

John Campbell is a professor of philosophy at the University of California, Berkeley. He has held Guggenheim and NEH Fellowships and served as president of the European Society for Philosophy and Psychology. He is a former fellow of the Center for Advanced Study in the Behavioral Sciences at Stanford University, Wilde Professor of Mental Philosophy at the University of Oxford, Professorial Fellow of Corpus Christi College, Oxford, and British Academy Research Reader. Campbell was awarded the Jean Nicod Prize in 2017.

Robert MacCoun is a social psychologist, a professor of law, and senior fellow at the Freeman Spogli Institute for International Studies at Stanford University. From 1986 to 1993 he was a behavioral scientist at the RAND Corporation, and from 1993 to 2014 he was a Professor of Public Policy and Law at the University of California at Berkeley. In 2019, he received the James McKeen Cattell Fellow Award of the Association for Psychological Science for lifetime contributions to psychology.